I0446480

Scientific Advice Meetings

A Guide to Successful Interactions with FDA, EMA and Beyond

Nathan Martinsberg and CF Harrison

ALSO BY THE AUTHORS

Scientific Due Diligence: A Guide for Investors and Investigators

Starting out in the Pharma Industry: Essential Knowledge for Life Scientists

Pharmaceutical Regulatory Affairs: An Introduction for Life Scientists

Aseptic Production: An Introduction for Life Scientists

From Test Tubes to Tonnes: Commercial Drug Process Development for Life Scientists

FOREWORD

Every drug development program has problems. But sometimes more difficult challenges arise. Ones which require discussion with regulatory authorities such as FDA to solve.

There is one chance to get this right.

One chance to provide your detailed, scientific arguments to the authority and sway them to your point of view. Success means the development program will continue on down the path of your choice. Failure can lead to cost overruns, delays or even cancellation of the project.

This book will:

- **Provide a step-by-step guide**: With in-depth discussion of each part of the process, from initial idea generation through to post-meeting stakeholder management, you can ensure that nothing is missed.

- **Help determine your ideal strategy**: Via a comprehensive guide to developing a formalised statement of what you want to achieve, perfect for keeping the team on track and getting management buy-in.

- **Encourage persuasive writing**: Information on structuring a briefing book, developing your chain of logic, and the art of writing ensures that you will maximise your chances of success.

- **Help new and experienced professionals**: Providing baseline knowledge for those new to the field as well as tips and tricks for late-career experts.

TABLE OF CONTENTS

INTRODUCTION

This book will guide you through the intricacies of scientific advice meetings. These meetings, held between a pharmaceutical development company and a regulatory authority, are critical for obtaining feedback throughout the drug development process.

The value of an advice meeting is directly proportional to the efforts you bring in. Well-established reasoning can sway a health authority to your side, while thoughtless questions can lead to devastatingly expensive changes in your plans.

We've written this book for the entire spectrum of experience – for those who are new to scientific advice entirely as well as for those who are long-time regulatory affairs managers. It can be read cover-to-cover, or you can jump to the most relevant section and work through from there.

So what do we discuss?

Basic knowledge

The initial sections describe the baseline knowledge which you need to have prior to any thoughts of requesting a meeting.

First we help you to understand the thinking of health authorities such as the FDA – their goals, their hopes for an advice meeting, and the typical reasons why your briefing book will fail. These are covered in Chapter 1 (page 12).

We then discuss several of the more important health authorities and their interaction pathways. You can find information on meeting FDA, EMA, as well as the Japanese PMDA and Chinese NMPA throughout Chapter 2 (page 18).

Following this we move onto the meeting process itself, focusing on the key stages throughout the journey.

The key stages of the meeting process

Deciding on your strategy

The first step in the whole process is to decide on your strategy. In other words, what do you really want to get out of the meeting. Answers to a specific question? Advice on your development plans? A friendly discussion on how to solve a problem which is coming up? All of these are possible, but you need to know which one you are chasing after.

A good way to determine this is by creating a **strategy document**, a formal record of your question, company position, and supporting arguments. The process of creating this is covered in Chapter 3 (page 33).

The initial application

Now you are ready to write the **initial application**. This is a very high-level overview of the problems which you are facing along with a general description of the drug development so far.

The main purpose is to allow the authority to decide if they want to talk to you or not. If you have a problem which requires discussion to solve, then they will grant a meeting (or some form of feedback, at least). If it is simple and clearly answered in published guidance documents then you will likely get a rejection.

An overview of the requirements for initial applications are shown in Chapter 4 (page 46).

The Briefing Book

The **briefing book** is possibly the most important part of the entire scientific advice meeting process. It lays out your arguments, provides the data to make your point, and persuades the health authority in question that your approach is the best one.

Because it is so important, writing a briefing book takes a lot of time. It needs to be clear, focused, and simple to understand – while also being persuasive and logically complete.

We discuss best practices for planning and writing a briefing book in Chapter 5 (page 60).

Preparation for the meeting

The briefing book has been submitted, which means you now need to prepare for the meeting and discussion.

There are several parts to this. You need to determine the right people to attend the meeting. You need to find the ideal response to questions which will come up during the discussion. And you need to practice beforehand, both your responses to questions and your approach when challenged.

We cover all of these topics in Chapter 6 (page 77).

Preliminary feedback

You will receive **preliminary feedback** several days prior to the meeting. This can be good, bad, inconclusive, or opaque. No matter how it looks, you and the team will need to assess the feedback and determine your response. Will you push back? Accept? Argue? There will be a lot of discussion at this point regarding the correct approach to take.

The required steps and best practices are discussed in Chapter 7 (page 89).

The meeting itself

The big day has arrived! Success at this stage is dependent on all the efforts performed in the previous stages, so hopefully everything has gone well so far.

Of course even the most-prepared will still run into problems. Thus Chapter 8 (page 104) describes the typical scientific advice meeting, the actions you should take, and how to keep the meeting running smoothly.

After the Meeting

But the work doesn't stop once the meeting is complete. There are meeting minutes to write, stakeholders to manage, and nerves to settle as you wait for the final feedback. This can be an intense period, sometimes even more stressful than the lead-up to the meeting itself.

Chapter 9 (page 115) will guide you through these stages in a (probably fruitless) attempt to reduce your stress levels.

And some extra information as well

In addition to providing information on the scientific advice process, we include several appendices which focus on necessary skills. These include:

- **Tips for overall writing skills**: Because writing well is the difference between a convincing briefing book and a meandering slab of text. Turn to Appendix 1 (page 121) for more information.

- **Review best practices**: Help the reviewers to focus their attention on the important topics rather than being bogged down in minor and pointless edits. These are covered in Appendix 2 (page 141).

- **Strategies for problem solving**: Before you can discuss a possible solution, you need to have identified a possible solution – and this can be a very tricky problem indeed. In Appendix 3 (page 170) we discuss some strategies which can help you identify the best solution to your problem.

CHAPTER 1: UNDERSTANDING THE HEALTH AUTHORITY

Scientific advice meetings are an excellent way to discuss problems in your development with health authority representatives. But getting full value from discussion requires you to understand what your options are and what those representatives will be looking for.

This section provides you with background information on the role which health authorities play and their general expectations.

How do health authorities see their role?

Health Authorities see themselves as *protectors* of public health. They have a responsibility to keep the population safe and those working there strongly identify with this role. Their scope can vary (FDA, for example, covers food, medicine, and various other products such as cosmetics and tobacco) but the underlying aim is to ensure that the public can trust the products which are available to them.

From a drug development point of view, they ensure that medical products reaching the market are safe and effective. At the same time, they also want to see newer, better medicines available to the population as qiuckly as possible. There is thus a constant balance between 'innovative' and 'well-proven'.

Importantly, they are focused on the **scientific evidence** which supports a potential medicine. Market size and commercial success are irrelevant, launch timelines are only important as far as they make a useful medicine available to the public. Thus you must ensure that your discussion and arguments are based on scientific principles.

They are usually very large and bureaucratic departments – FDA, for example, has around 20,000 employees. However the sheer number of areas they need to cover means that they need to specialise and you will often have repeated interactions with the same people. This means that your reputation from one meeting can easily influence your outcomes in a later one.

Although they seem formidable, health authorities *want* to see effective medicines available to their people. Thus they are willing to support drug development to ensure it is done *correctly* and in a manner which is most likely to support approval. This support comes in many forms, but the most useful support is direct feedback on specific topics of your drug development. And to do this, they provide scientific advice meetings.

What is a scientific advice meeting?

A scientific advice meeting is one where the company developing a drug (the **applicant**) meets with representatives of a health authority to ask for advice on certain topics. It is not a coaching session or general discussion – they expect that you have the experience to do most of the development yourself. Instead each meeting focuses on providing *answers to specific questions* asked by the applicant.

These questions can vary immensely and cover almost every facet of drug development. They can be quite broad and ask for wide-ranging advice, as typically seen in early development. Or they can be laser-focused on an obscure yet vital topic which threatens to derail the entire program (which occurs more often than you would expect in late-phase).

There are a number of different meeting options available and we cover some of these in Chapter 2 (page 18). Most companies will focus on discussion with the FDA (as the USA is a major market for pharmaceuticals and the FDA has very high requirements for approval). However health authorities from other countries can also provide vital information regarding their specific regions. Scientific advice meetings should therefore be an important part of any global development planning.

Although we use 'meeting' throughout this book as a general term, you should keep in mind that this doesn't just cover a standard face-to-face meeting. The FDA, for example, considers that video- or tele-conferences as well as simple written feedback to be different facets of a 'formal meeting'.

Who is the health authority reviewer?

It is important to remember that the most important reader of your briefing book is *not* the highest member of management who takes a look. It is the reader at the health authority, whether it be FDA, EMA, or anywhere else, who must be persuaded by your arguments.

The typical health authority reader is clever and an expert in their particular branch of the pharmaceutical world. They are concerned about patient safety and the quality of the medicines which are provided to their fellow citizens. They see questions and applications from many different companies on many different topics, all of which will impact their thinking when they respond to your briefing book. This is why advice meetings often go down unexpected paths – they have seen other approaches for similar problems and are trying to direct you towards them.

Although they are focused on safety and efficacy, they are *not the enemy* and they are not *actively opposed* to you and your suggestions. Scientific advice meetings can range from collaborative to confrontational, this is heavily dependent on the topic being discussed and the data which you are showing. But this all comes from a *scientific* opposition, you will not find a reviewer turning hostile simply because they dislike something your company has done.

Regulatory affairs managers in a pharma company rarely have close personal contact with their counterparts in the health authority, at least at the lower levels of the hierarchy. Thus they tend to focus on 'organisational knowledge' of what the reviewer is looking for – i.e. we sent off a briefing book last year, and they asked for this, so we need to ensure it's included this time too. This can be useful (you gain knowledge over multiple interactions) but also a danger (as you can pick up bad habits too).

It may be challenging to understand the particular reviewer handling your case, but there are a few basic attributes which will generally hold true:

- **They are judging you**: They aren't particularly interested in *learning* something new from your briefing book or your particular challenge. Instead they are reading to judge your arguments and decide if they are sufficient for their requirements.
- **They are picky**: They actively look for holes in your data or arguments. Weak data, logic gaps, overblown claims, bad statistics – areas where they believe you could (and should) do better. They don't simply read to 'follow along' with your proposals.
- **They are in a rush**: There is never enough time, no matter where you are. The regulatory reader will be trying to make a decision quickly so they can get onto their next problem.
- **They jump around**: Some may read from the start of the document to the end, but most will jump around to different paces, skimming some sections and going into

more detail on others. They also use search tools or hyperlinks to move between sections to find specific information. You can't predict where they will do this, so each part of your document needs to stand on its own.

Why did your last briefing book fail?

It is difficult to properly define 'failure' because there is a very wide range between an unsatisfying HA interaction and an outright rejection in the initial screening phases. But you can take the idea of failure in the broadest sense – your document has failed if it doesn't persuade the reader that the argument it makes is correct.

How does this happen? There are many reasons which can cause a briefing book to fail, and you've likely seen it happen before:

- **Not explaining**: One of the biggest mistakes is simply dropping results into the document without correctly explaining the context or significance. The reader is very unlikely to have the same level of background knowledge that you do, so you need to explain everything, regardless of how 'obvious' it seems.
- **Hiding problems**: It is tempting to gloss over problems or weaknesses in the data, hoping that the reviewer simply won't realise. This rarely works (they aren't stupid after all) and will just lead to a greater focus on those areas. It is almost always better to acknowledge weak points and discuss them.
- **Leaps of logic**: A briefing book needs to present an unbroken chain of logic that leads the reader from the data to the final conclusion. Breaking this chain by not explaining steps correctly or making it impossible to follow will confuse the reader. A confused reader will then make up their own mind, rather than following yours.

- **Badly presented data**: Your data needs to be presented in a way that is designed with the reviewer in mind. Anything you show should clearly fit into the narrative, and it needs to be easy to find critical data. Long appendices of values are necessary sometimes, but you also need sensible summary tables as well.
- **Information is hidden away**: This is not deliberate hiding of issues, but rather that the briefing book is written in such a way that the reviewer can't find the information they need. A typical health authority reviewer sees many different document formats and constructions over the course of a year – they will not remember how yours is set out either. A good briefing book needs to make key information easily available to the busy and distracted reader.

A regulatory document needs to be *usable*. It should help the reader to understand the argument being presented and the underlying facts. Thus whoever plans, writes or reviews needs to keep the following in mind:

- Who is my target reader, and when will they be reading this?

- How will they use it?

- What questions are we answering for them?

- And how does it connect to other work which we've previously done, but which we aren't showing in this document?

If you can keep all of these in mind, you are well on the way to success.

CHAPTER 2: ADVICE MEETINGS ACROSS THE GLOBE

The world is large and there are many different health authorities which will have a say in your drug development and approval. In this section we cover some of the major ones – the US Food and Drug Administration, the European Medicines Agency, the Pharmaceuticals and Medical Devices Agency from Japan, and the Chinese National Medical Products Administration.

Advice meetings in the United States of America (FDA)

Scientific advice meetings are known as **formal meetings** by the FDA. They are regulated under the charmingly-named **PDUFA, BsUFA and GDUFA** regulations of the United States.

These acronyms stand for the Prescription Drug User Fee Act, Biosimilar User Fee Act and Generic Drug User Fee Act. The laws create a number of conditions and timelines for the FDA to act upon. As part of this, they also include many requirements for the scientific advice meeting process.

Available meeting formats

There are several types of meetings which an occur, ranging from relaxed teleconference discussions with the procedure manager through to stressful Advisory Committee meetings (in which company attendees rehearse for months beforehand to ensure that they cover all the relevant points). In general, however, you will be taking part in one of the 'typical' formal meetings with the FDA.

The meetings themselves can be held in several different ways, namely face-to-face, via tele-/videoconference, or as a Written Response Only (WRO).

The most traditional of these options is the **face-to-face** meeting, where a team of experts from the pharmaceutical company will dress up in suits, head out to the FDA campus, and meet with a team of experts from the Agency. A second option is the **teleconference or videoconference**, where all participants dial in from a remote location. This became very common in the post-Corona world and you can expect many meetings to be virtual going forward. Technology is still a challenge, however, and the FDA seems to prefer teleconferences as there is less pressure on the internet connection (particularly when the applicant is located overseas).

Finally a **Written Response Only (WRO)** occurs when the topic is considered to be answerable within a feedback document. This often occurs when the FDA considers your questions reasonable, but not important or complex enough to take to a full meeting. Once they have decided then no meeting will occur, you will receive a document outlining the FDA response to your questions. It is possible for the applicant to request a WRO but it is generally not a good idea – FDA feedback is then final and you have no further opportunity to persuade them of your viewpoint.

Meetings under PDUFA

There are four general *types* of meeting which are possible under PDUFA, known as Type A, Type B, Type C, and (just to break the pattern) Type B (End of Phase). All of these have different goals and thus are best suited for certain problems which you might have.

- **Type A Meetings:** Are designed to solve roadblocks, they focus on getting your project moving after it has hit a regulatory halt. The most common reasons to call for a Type A meeting are after a clinical hold (when your clinical trial has been stopped by regulators), after a complete response letter (when your application has been rejected) or after a refuse-to-file (when they didn't even want to look at the application). In other words, the FDA has done something you don't like, and you have to find a way around it. As these tend to be difficult meetings, the FDA prefers that you speak to them first before requesting one.

- **Type B Meetings:** Are milestone meetings occurring when a major point in the development has been reached. Important examples here are the pre-IND or pre-BLA/NDA meetings, or as a meeting prior to applying for emergency use authorisation. You may also call for a Type B meeting if there is discussion on risk management or post-marketing commitments that doesn't fit into a normal dossier review cycle. Generally if you have a problem pop up during development, you will end up in a Type B meeting.

- **Type B (End of Phase) Meetings:** Occur at the end of a clinical trial period (e.g. the end of Phase I studies) and prior to moving into the next trial phase. They are a chance to discuss clinical results and remove any concerns the FDA might have regarding your safety or efficacy results at this early stage.

- **Type C Meetings:** Are essentially any other meeting which doesn't fall under the other categories but which still relates to development and review of a medical product. Beyond this Type C meetings are used if your company would like to introduce a new biomarker as a surrogate endpoint for the clinical trial.

The PDUFA legislation sets a number of goals for the FDA in order to ensure that they respond within the appropriate amount of time. This is not always achieved, particularly in periods when there is a lot of additional work (the COVID epidemic is a good example). However they usually hold to the timelines shown in the following table, which lists the time from meeting request going in to the Agency until the next milestone. As with all regulatory timelines, it is counted in *calendar days*, not working days.

Standard timelines for meetings under PDUFA

Meeting type	Initial response time [days]	Meeting package required	Meeting scheduled within [days]	Written Response Only within [days]
Type A	14	With request	30	30
Type B	21	>30 days prior meeting	60	60
Type B (EOP)	14	>50 days prior meeting	70	70
Type C	21	>47 days prior meeting	75	75

Meetings under BsUFA

There are also meetings which are specific for development of biosimilars. Regulated under BsUFA, these are the BIA and BPD Type 1 – 4 meetings.

- **Biosimilar Initial Advisory (BIA) Meeting**: This is an early assessment meeting, basically getting the FDA's opinion as to whether a biosimilar license would be theoretically feasible. There is very little data at this point and the FDA themselves only want to see a small set of results. Anything more complex will require one of the other meeting types.

- **Biosimilar Biological Product Development (BPD) Type 1**: This is essentially a Type A meeting in a biosimilar coat. It's designed for discussing stalled development (clinical holds, complete response letters, refuse-to-file letters) and important safety issues.

- **BPD Type 2**: Are intended for discussion of a specific issue or to request guidance on a very targeted topic. The FDA will review the data which you provide them as part of answering the question, but it has to be focused on your problem – general reviews of 'everything' are off the table.

- **BPD Type 3**: These meetings cover broad topics such as 'are we a biosimilar?' or 'is this clinical study report sufficient?'. The FDA will take the time to look through large amounts of data and provide a thorough response with detailed advice.

- **BPD Type 4**: Type 4 meetings are pre-submission meetings, providing the reviewers with information on the dossier structure, key studies, and any issues they may need to keep in mind. This is basically the final chat before the dossier goes in.

Generally you can expect these meeting requests to be handled within the timelines shown in the following table. Note that all times are counted in *calendar days*, counting from receipt of meeting request and package.

Standard timelines for meetings under BSUFA

Meeting type	Initial response time [days]	Meeting scheduled within [days]	Written Response Only within [days]
BIA	21	75	75
BPD 1	14	30	N/A
BPD 2	21	90	90
BPD 3	21	120	N/A
BPD 4	21	60	N/A

Meetings under GDUFA

The GDUFA system is different to that of PDUFA, reflecting that fact that generic development has an alternative focus (i.e. creating a safe and reliable copy of an approved drug, rather than showing efficacy of a new one).

It also includes a number of additional meeting types for **complex products**. The FDA defines these as products which have complex APIs (such as peptides), formulations (liposomes, etc.), routes of delivery (dermatological patches, etc), dosage forms (inhalers, etc.) or where there is a complicated drug/device setup (such as pre-filled auto-injectors). As these are more difficult to develop than the usual generic, the FDA tries to support the process with additional meeting types.

Available meeting types include:

- **Product Development Meeting**: Targeted at complex products, this allows the applicant to discuss specific issues in an *ongoing* development program. This means that the FDA expects you will have started development and have data to discuss, it's not just for fun. They also try to limit applicants to one meeting per year, though it can be more if there is a true need.

- **Pre-submission Meeting**: Intended to occur 6-8 months before your planned submission date, presenting unique or novel data which will be part of the ANDA and get FDA feedback on it. This should generally follow after at least one product development meeting and the FDA counterparts will often be present at both.

- **Mid-cycle review meetings**: Occur during the review of the drug application, appropriately enough right after the mid-cycle review letters have been received. This is a very limited meeting and you are only allowed to ask about the FDA assessment of your data, no new information can be presented at this time.

- **Enhanced mid-cycle review meetings**: Are, as the name suggests, a better version of the normal mid-cycle. In these FDA allows you to discuss new data or scientific approaches which you may want to take in your development process, though they won't provide an in-depth assessment of anything new at the meeting.

- **Post-CRL meetings**: Occur once you have received a **complete response letter**, i.e. a rejection. This allows you to discuss the gaps in your application as seen by the FDA; and how they feel you could solve these problems.

Standard timelines for meetings under GDUFA

Meeting type	Initial response time [days]	Meeting scheduled within [days]
Product Development	14	120
Pre-submission	30	60
Mid-cycle	N/A	30
Enhanced mid-cycle	N/A	90
Post-CRL	14	90

If the FDA decides that your meeting request is not worth granting (i.e. there is not enough data, the answer is covered in guidance already, etc.) then they will push you towards the **controlled correspondence** pathway, a method for generics manufacturers to make a written request for feedback. As always, the decision on whether to take up a meeting or not is up to the FDA.

Advice meetings in the European Union (EMA)

EMA is simultaneously simpler and more bureaucratic than the FDA in the way they handle advice meetings. The meeting options you have are straightforward and simple enough to request. However the process after that quickly becomes complex and extremely inflexible – EMA has a process, and they will stick to it no matter what.

On the plus side, organisations developing medicines can request a scientific advice meeting with EMA at *any* stage during development. This applies regardless of whether the drug will be approved by EMA or not.

The meeting and advice formats are named:

- **Scientific advice meetings:** Allow you to ask EMA for their opinion on specifically-highlighted questions regarding your development process.

- **Tailored advice meetings:** Are specific to certain development programs. This is possible for biosimilar developers, but also for academics or not-for-profits who are attempting to re-purpose existing drugs for a new indication.

- **Protocol assistance:** Specialised advice for those developing an 'orphan drug', one intended for a small but as-yet untreatable portion of the population.

Scientific advice meetings

A scientific advice meeting can be requested at any time, no matter if it is during development (prior to submission of your application) or afterwards (in the post-authorisation phase).

EMA is fairly clear in stating that advice meetings should be for topics where there is no relevant **guidance documents** or Pharmacopoeia monograph. As EMA is quite prolific when it comes to writing things, they assume that only the really tricky questions will fall through the paperwork cracks and come to a meeting. This is not actually true and you can ask for a meeting on basically every subject, with a subsequently higher risk of EMA rejecting your meeting request.

This requirement for 'tricky questions' is loosened for those who have limited experience in developing medicines. In these cases, usually academics and small-medium enterprises, EMA will provide additional help for them to have a better chance of success. There is also additional financial and bureaucratic support for these 'newbies' to the field.

Because of the broad scope in EMA scientific advice meetings, you can ask questions about many different aspects of your development program. Some typical examples:

- **Development strategy** such as bridging approaches for generics / biosimilars, safety database strategies, paediatric development and orphan designation.

- **Quality aspects** such as manufacturing, controls, and analytical testing of your medicine.

- **Non-clinical aspects** such as your choice of toxicological and pharmacological tests or study design.

- **Clinical aspects** including the choice and design of studies and endpoints, risk management plans, and other post-authorisation measures.

- **Methodological issues** regarding the choice of statistical tests for your data, data analysis, and methods used in modelling and simulation.

There are only a couple of areas which EMA will simply refuse to answer within these meetings. The main one is acceptability ("are these specifications acceptable?" - the answer here is inevitably 'this is a review issue'). The other is topics relating to studies and outcomes in children – these are referred to the Paediatric Committee.

Do be aware that EMA has a strong commitment to open information in the regulatory process. The questions you ask won't be publicly available at the time of meeting, but the final post-approval assessment report (which *is* public) will include a summary of your questions and whether or not you followed their guidance. So be careful with what you request and how it will look to the outside world.

Timelines

Timelines for EMA are slightly different to FDA. While FDA commits to respond within X days after submission, EMA runs on a fixed schedule with pre-published slots for each activity. That means that if you have a topic in, say, January 2024, you need to have the briefing book in place by Jan 31st. Any later, that's too bad, you have to wait a month until the next slot.

Once your briefing book is submitted then EMA will take it up for discussion a few weeks later within the Scientific Advice Working Party (SAWP). Depending on their assessment you will either receive a written feedback around day 40 post-application, or they will invite you to a discussion meeting. The meeting occurs around 2 months after the start of the process, again in line with the published schedule. Final feedback then comes on Day 70.

Tailored scientific advice meetings for biosimilars

Tailored scientific advice is available for biosimilar manufacturers. The aim here is to provide detailed feedback on study design based on the data which they have already developed.

These meetings allow the applicant to ask more specific questions about biosimilar development – about the suitability of the data package, about differences seen to the originator, about the clinical component, etc. Due to the complexity they take about one month longer than the standard pathway, and thus you can expect the final EMA letter around Day 70 or Day 100.

The Parallel Scientific Advice process between EMA and FDA

EMA and FDA have a program known as Parallel Scientific Advice (PSA), which allows both regulatory agencies to provide feedback on a topic at the same time. This is intended to avoid the back-and-forth which occurs when trying to balance differing feedback from the two.

The PSA is intended for important medicines which are being developed in areas where guidelines don't exist or where they differ between the two agencies.

Many drugs will be aimed at both markets and you will often receive contradictory advice from FDA vs EMA when doing separate meetings (particularly on the design of your clinical trial, though there are many other tricky areas). As such it's worth trying to get one of these if you can.

Timetable for EMA / FDA parallel scientific advice

Time	FDA	EMA
Anytime	Submit request for PSA to both agencies	
Day 0	EMA and FDA validate meeting package	
Day 15-25	FDA internal meeting	EMA SAWP discussion
Day 30-34	Agencies send list of concerns / issues to each other	
Day 35	FDA and EMA meet to discuss	

Day 65	Meeting between EMA, FDA and the sponsor	
Day 75-95	FDA meeting minutes 30 days after meeting	EMA final advice 10 days after meeting

Advice meetings in Japan (PMDA)

Scientific advice meetings offered by PMDA are referred to as *consultations*, and they are offered for multiple aspects of the development process. It should be noted that the meetings are held in Japanese, though translators are of course allowed to attend.

One of the major areas in which PMDA provides consultation is that of clinical trials. Discussion can be held before each trial phase to focus on applicability of the study and endpoints. PMDA tries to ensure that the Japanese population is sufficiently represented in clinical trials, their goal is to ensure the drug is safe and effective for Japanese nationals. This can be very, very difficult to work around and so it is always worth meeting PMDA if you are intending to market in Japan at some point.

Beyond that you can ask questions relating to the typical topics (pre-clinical, CMC, etc). In addition PMDA offers 'Prior Assessment Consultations'. This is in effect a pre-submission meeting, where the data which will be provided in the application is evaluated and a list of issues provided.

Small start-ups and academics are also able to use the 'Regulatory Science Consultation' pathway. These provide advice on the studies and clinical trials which are needed for early development, but tailoring the discussion for applicants with limited experience in drug discovery. Fees are also waived if you can meet the prerequisites, which basically involve your company to be small and not overly wealthy.

Standard timelines for a Formal Consultation with PMDA

Time relative to meeting date	Activity
2-3 M prior	Applicant submits meeting request PMDA/Applicant agree on meeting date within 5 days
1-2 M prior	Payment of fees and 'official' meeting request
5-6 wk prior	Briefing book submitted and 'discussion period' begins During this time PMDA will ask several rounds of questions to be answered by applicant, before finally providing an initial opinion 5 days prior to the meeting
Day 0	Face to face meeting is held
1 M later	Meeting minutes will be provided by PMDA

Advice meetings in China (NMPA)

Scientific advice meetings can also be held with the Chinese health authority, NMPA. Meetings are held in Mandarin Chinese, though translators are also welcome to attend.

There are three main categories of meetings which are possible with NMPA

- **Category I** meetings are intended to address major safety issues during a clinical trial or major technical issues during development. Meetings will be scheduled 30 days after the request, and the briefing book needs to go in at the same time to hold to this timeline.

- **Category II** meetings are held during drug development at milestones such as before Phase I, prior to Phase III, or directly before NDA submission. These occur 60 days after the request, with the briefing book required 30 days prior to the meeting.

- **Category III** meetings are for everything that doesn't fall under the other meeting categories. The meeting will occur roughly 75 days after request, again with the briefing book being required 30 days prior.

In all cases a preliminary opinion will be provided by NMPA 2 days prior to the meeting. At this point you can change to a written response (assuming you are happy with all of the feedback), though this is rarely a good idea. The meeting will last 1-1.5 hours and you can expect the final minutes around 30 days afterwards.

CHAPTER 3: ALIGNING ON STRATEGY

This is not always obvious to everyone on the project team, but yes, you should have a strategy in place before you go for a scientific advice meeting.

Why?

Well, it's pretty simple actually. You request a meeting when you need help on a topic, usually an issue which you are dealing with. There are inevitably several ways in which this issue can be solved, one of them will presumably be better from your perspective than the others. You just need the health authority to agree that this is the case.

How do you get them to agree? You need a strategy. And that is what this chapter is all about.

Why is that so difficult?

Drug development is an area characterised by large projects involving many different experts. As you would expect, they also have problems. Large ones, small ones. But the problems which require a scientific advice meeting are usually major ones affecting multiple areas of development, and so the briefing book will be prepared by representatives from clinical, engineering, analytics, pharmacology, and more. A successful briefing book and advice meeting must bring the knowledge of these disparate experts together into one focused argument.

The important word here is *focused*. Information should be merged in a way which *clearly and consistently* supports a set of key messages and issues. Successful meetings need a coherent and aligned approach which creates a strong logical trail for the Health Authority to follow, one which leads them towards *your* desired conclusion.

In practice this is a challenge. The successful team will need to overcome messy data, limited writing skills, changing opinions on strategy, and complex document structures. We describe these these problems further in the following sections.

Messy data

Data from pharmaceutical development is often 'messy', there are gaps or sections which are difficult to fully explain. Gaps can occur anywhere, be it from clinical side (e.g. extrapolation of surrogate markers to progression of a disease), or from CMC (e.g. differences in quality attributes following a tech transfer). Yet the team needs to bring this imperfect data together in a way which supports complex scientific arguments.

A strong argument despite messy data is possible, but requires the input of many experts in different domains. This is the classical 'cross-functional team', an excellent idea in theory but one which often runs into problems in practice. It is rare to find a team which instantly agrees on interpretation of data or the best way forward, and even rarer to find a team who can *articulate* this interpretation.

Given this, you need time to find the correct (and most persuasive) interpretation of all data. Doing this while writing the briefing book is a recipe for disaster. Instead, these discussions should happen from an early stage, with the strategy document serving as an evolving record of the current team thinking. Gaps and problems can be highlighted for further work and assessment, and the target approach can be changed if new information comes in.

Limited writing skills

Briefing books usually have a 'main' writer (generally someone from regulatory or clinical) with support from the team on sections which match their area of expertise. Multiple people will be involved in writing, even more will have helped to create the relevant source documents. Very few will have formal training in writing. You are most likely to have a team of experts with no interest in writing the briefing book, but who have nonetheless been forced into doing so.

This is made worse by the long drug development process (6-10 years) and the scattering of scientific advice meetings across the entire development timeline. Team members will change several times over this period, which means the writing process is often unfamiliar for those who are responsible 'this time around'.

Because it is unfamiliar the authors will write based on their experience and training. Most people in the field have scientific backgrounds and so default to the 'scientist' approach – experiments are done, results are analysed, the paper then describes what happened. The data will be laid out objectively with minimal interpretation, using a 'sequential' approach to describing the results (we found this, and then this, and then that). Then the authors will summarise the results in isolation rather than within the context of the entire development program.

This is unfortunate when found in an unimportant source document, but it is *dangerous* when it occurs in a briefing book. Your goal is to push the reader towards your point of view, and a dry restatement of results (while scientifically accurate) will leave too much room for their interpretation. This significantly increases the chances of getting feedback which you don't want to hear.

You can minimise the impact by creating a clear strategy document which outlines both the approach to be taken and the application of this in the briefing book. The pre-aligned layout and logic trail ensures that the final document will be as persuasive as possible.

Changing strategic opinions

Scientific advice meetings are important. Feedback from the Health Authority can completely change your development program, timelines, or overall chances of success. As this obviously has major implications for the company, it is no surprise that higher management wants to be involved in the process.

And of course no self-respecting middle-management function would *ever* allow a document to go to upper management without checking it first. Which means that any briefing book will inevitably have multiple review rounds – within the author group, within their direct management circle, and within the higher levels of management as well.

Every review round brings in more opinions, often conflicting, often changing from round to round. Thus you will end up rewriting large amounts of the briefing book in a mad rush just prior to submission deadline because one senior manager made a last-minute fuss and absolutely insisted on changing this one particular part and then another wanted a second change while a third... You get the idea.

To avoid this, we create a strategy document and *get buy-in of the strategy* from all the involved stakeholders *before* any major writing happens. This ensures that everyone is clear on the approach, and reduces the chance that major changes will be requested during later stages.

Complex document structure

Any briefing book for a scientific advice meeting tends to be a long and complex creation. In general the briefing book will follow a fairly standard layout, usually defined by the HA themselves, however there is enough flexibility in the rules that no two documents will be set up exactly the same. Similarly there is usually a split of details and summaries between different document sections or appendices, all of which needs to be internally consistent. You also need to clearly restate the intended message at all levels of detail.

Defining the overall strategy and implementation approach first makes it significantly easier to create the briefing book. The main questions are already answered at the strategic level – what information goes where? And how should we argue our point? Thus the writers can quickly create the body of the document with minimal back-and-forth.

The Solution to these problems – a Strategy Document

As described above, there are many pitfalls involved in crafting a clear argument for your upcoming scientific advice meeting. The best way to avoid these is to define *and document* the overall strategy at the start of the process.

The Strategy Document, sometimes described as a 'Seed document' or 'Action Plan', allows everyone to add their input and ideas *at the start*, before any significant drafting is performed. Strategy documents also provide a quick overview of the major logic and risks involved, and are thus useful for discussing the regulatory strategy with management.

The following sections describe the contents of these documents in more detail. However the core of any such strategic document is the list of issues which will be addressed with the health authority. Specifically, it will state the issue (the problem you are having), the company position (your proposed answer), the key logic supporting that position (why do you think we are correct) and the available data to support that logic (to prove that our answer is the right one).

The process by which this list of issues is created forces the team to work together, then defend their reasoning to different layers of review. This ensures high-quality arguments and puts everyone on the same level of understanding before briefing book preparation can begin.

So before we begin all of this – what does a strategy document require? Are we spending months of valuable time on a pointless task?

Well, no. Partly because it forces you to plan out the writing process, which you need to do anyway. And partly because it doesn't take that much time. At least, if you can keep all of the reviewers on track...

Planning the writing process

Writing a *good* briefing book requires time – time to get the data in order, to decide how you will present your argument, time to write and time to get feedback. Each step is time-consuming yet vital for the final product, and everyone involved will *consistently underestimate* the time and budget costs required to finish. Which inevitably leads to the oh-so-typical mad rush to get it done in the week before the deadline.

To avoid this you should always work according to a plan, an organised timeline of how the briefing book will be drafted, reviewed, and finalised. This is essential for preventing late delivery of drafts, hurried review rounds, and subsequent incomprehensible documents.

The plan should include all of the major milestones (writing the first draft, review, etc.). Generally you will set up the plan by working backwards from the final due date, the point at which the regulatory submission will occur. You need to be careful to allocate the required amount of time for the 'forgotten milestones' of review, revision drafts, and the all-important publishing and signature rounds.

If multiple documents are being prepared then you can often juggle due dates around for maximum efficiency. In particular you can try to place a hold on documents which need strategic input or information from other documents, allowing you to push through to finalisation with the simpler, descriptive reports.

Planning is also a good moment to decide which experts should be performing review at each stage – often some groups will take longer or will provide more important comments than others. These will usually lie on the critical path for document review and so will need more attention during the process.

The *ideal* document plan will have both best-case and base-case assumptions. This allows you to adjust your planning in case of delays or problems (and there are always delays or problems). Having said that, creating these additional timelines is a lot of work in practice and is usually overkill for the majority of documents.

A disadvantage of planning out the process like this is the sudden discovery that writing a briefing book is actually quite expensive – the time and salaries of those involved usually makes for quite a bill. Despite this it is immensely preferable to have these costs out in the open rather than hidden away as a random extra budget expense at the end of the month. Similarly openness on costs helps to slim down the review process – nothing quite encourages efficient review like telling people that each round costs several thousand dollars.

Sounds great, right? Just remember that documents plans suffer from the same problem as all planning does – sometimes reality does not match up to your hopes or requirements. Take a deep breath and remind yourself that even a slightly wonky plan is better than no plan at all.

Typical strategy document timelines

Obviously every project is different and there is no way to clearly say 'this is what you should be working towards'. But there are some general timelines which many strategy documents can follow.

- **Writing the first list**: Will involve many disciplines as they come together and brainstorm ideas for questions. Generally many of these will have been brought up by the SMEs beforehand, but you can expect a lot of discussion as it comes together. Plan for at least two meetings of several hours each, preferably separated by a few days to a week to allow everything to be mulled over.
- **Creating a first draft**: At this point the selected issues are covered in more detail – summarised information and opinions will be placed into the table and modified after each meeting. You can expect at least a week for this stage, possibly more if there are many topics or underlying data is lacking.
- **Expanding into an understandable document**: The table from before is fleshed out, it now becomes something which people outside the group can read and understand. This will often run in parallel to the first draft and the speed depends on the level of detail provided in the prior step. You can usually expect at least a week here as well, often more if the teams' writing skills are less than they should be.
- **Review by the SMEs**: Now is a good chance to give the experts a break from their individual parts and ask them to review the strategy document as a whole. The team will be deeply involved in the topic at this stage so

should not need more than 2 days. Another 2 days to discuss and incorporate comments is also usually appropriate.

- **Review by your project group**: Every drug development will have a project team who are not directly involved in the strategy document preparation. Getting them to review is a good chance to see if the document works for 'not quite outside' readers, and they should focus on *clarity and completeness* of the arguments. Again, 2 days for review and 2 days to implement comments is a good mix.
- **Review by functional management**: Your experts have managers, and they will want to provide their opinions too. Sometimes they are even helpful. The functional managers should focus on the *clarity* of the arguments and the *strategy* proposed for their respective disciplines. This normally takes a bit longer as their calendars are more packed, so assume at least 3 days review + 3 days implementation.
- **Review by upper management**: Scientific advice meetings are important, they can completely change development programs, and so upper management will want to be involved. Their review should focus on the *overarching strategy* and should avoid getting stuck into the details (though this is difficult). As for functional management, you can expect at least 3+3 days.

At this point you should finally have an aligned, management-approved strategy document. Make sure you save it in a safe place, preferably signed and dated, and send copies of the final version to everyone involved. This will not only inform your approach for the later steps, but covers your behind when that approach isn't successful.

Typical timelines required to create a strategy document

Strategy Document Phase	Typical Duration [weeks]	Cumulative Time [weeks]
Writing initial list	1	1
Creating first draft	1 – 2	2 – 3
Expanding to readable document	1 – 2	3 – 5
Review by SMEs	0.5 – 1	3.5 – 6
Review by project group	0.5 – 1	4 – 7
Review by functional management	1	5 – 8
Review by upper management	1	6 – 9

Creating a Strategy document

So how do you go about creating a strategy document? The initial stages vary depending on your situation. If you are planning for a general advice meeting in which many parts of the program will be discussed, it is worth taking time to determine all the possible challenges. In this case, you can start from the beginning.

However, if you have a specific problem which you need to solve then you need to focus on this. Additional brainstorming is only helpful as it brings you towards solving this goal. As such, you can jump directly to Step 3.

Step 1: List all potential issues requiring feedback
The first step in the process is to create a list of project issues. Especially initially, this is a matter of the entire team thinking of as many possible risks and challenges as possible. They can do this in many ways, including:

- Sitting together and brainstorming for ideas and issues which may affect the project.
- Examining the competitors labelling and other public information to see what issues they may have had, and how they may have dealt with them.
- Each expert takes the viewpoint of their speciality, running through the entire development process and listing issues which might occur.

All of these issues and potential problems are written down and sorted, essentially making a giant list of the challenges the project could face. Writing these down immediately ensures that ideas won't be lost over time, and indeed it often helps to have a dedicated scribe during any brainstorming session who will catch any idea – no matter how far-fetched it may seem at the start.

After this initial brainstorming, it may be worth investing some time within the group in order to really nail down the best wording of these issues. This may occasionally seem like a waste of time, but saves a significant amount of stress later on when everyone is arguing about problem definitions.

Also remember that any ideas which pop up later in the project should also be included in this list – the document is intended to be a 'living' document with constant updates.

Step 2: Prioritise the issues

You have a long list of problems, which is great. But you only have one to two hours to discuss them, which means you need to prioritise.

Generally the project teams will have a good idea of what is highly critical and absolutely requires discussion, as well as what is relatively unimportant. This means you can quickly decide on the 'must-have' topics from your list. The challenge then comes with the next layer of topics, those which are 'mostly' important.

Why does this happen? The project team has representatives from many different functions with wildly different focuses. You may need to decide whether a issue with the drug manufacturing control strategy is more important than a question on the timing of launching a shared safety database. Which of these has priority? It is very, very difficult to say without looking into the issues and analysing what it means to the success of your project.

You can expect a lot of discussion at this point and it is very easy to fall into personal arguments and shouting. As such it's up to you to take a leadership role, keep the team on track and focus them on the problems. Otherwise you will find yourself going round and round in circles, deciding nothing.

Step 3: Decide on the question to be asked

This sounds easy, right? Well, it's not. It is surprisingly difficult to go from knowing you have an issue to knowing how to ask for help. As described in the following sections, there are a few things you need to keep in mind as you develop the questions for your main topics.

Focus on the future

Your questions should be forward-looking. Focus on the development strategies which you will use, either as a general request or as a specific question on your particular problem. In other words, you need to have an idea of what you are going to do, and then ask if this is appropriate.

Indicate what you plan to do

You can't turn up to a meeting and say 'what should I do?' (or, more realistically, 'What would be a suitable clinical study plan for this product?'). The health authorities (quite reasonably) expect that you will have done a pre-evaluation of your issues and determined some sort of approach to mitigate them. Failing to do this will net you a non-answer such as 'We recommend developing a plan in line with guidance X, Y and Z".

This means that even if you have no idea what you are doing (which is hopefully not the case) you should have a suggested solution. It may not be a *good* solution, but this is enough to lead into a discussion with the authority.

You cannot ask for pre-assessment

A trap which many fall into is asking in a way that requests pre-assessment of the data. What does this mean? The agency *cannot tell you if your data is acceptable for approval*. They make this decision once the full regulatory filing has been received, balancing all of the data available against the public benefit. Asking if your plan is appropriate is fine, asking if the data is sufficient is not.

In other words, you can ask "Is the overall development *plan* adequate to support marketing authorisation?" But you can't ask "Are the phase 3 study *results* adequate to support marketing authorisation?" Your question needs to focus on the approach which will be taken, not whether your data is sufficient.

Some example questions

This is fine, but it's much easier to design a question if you know what it should look like. So here are a couple of questions which EMA has already indicated are acceptable:

- Are the patients to be included in a study sufficiently representative of the population for whom the medicine is intended?

- Are the planned measures to assess the benefits of a medicine valid and relevant?

- Is the proposed plan to analyse results appropriate?

- Does the study last long enough and include enough patients to provide the necessary data for the benefit-risk assessment?

- Is the medicine being compared with an appropriate control?

- Are the plans to follow the long-term safety of the product appropriately designed?

Step 4: Initial draft of the strategy document

At this stage you have a prioritised list of issues and a rough idea of the question / proposal which you intend to make. That's great, you're halfway there! Now you need to formalise it.

In this stage the initial information is converted into a draft strategy document. There are several sections which are helpful to include in a 'good' document, though realistically the most important one is the list of issues and possible responses. Nonetheless we include some detail here to help with designing your templates (and because it is *much* easier to get through management reviews when this information is in place).

Meeting purpose

There are many different reasons to have a scientific advice meeting, but the questions will often revolve around one or two core purposes. It helps to explicitly state this in the strategy document to ensure everyone is on board. It also provides vital context for anyone outside the team and lacking the in-depth background.

Country scope

You may be aiming to discuss issues with the FDA, or with EMA, or even the Cuban health authority CECMED. Or all of them. The countries in scope affect the information you need to provide and the timelines you will work with, so this is important information to define up front.

Background

Here you will provide a high-end summary of the project (for those who are unfamiliar with the project), a more detailed look at the meeting scope, and a clear statement of your overall company goals. There is some overlap between this section and the meeting purpose section – it is best to think of this as providing depth and explanation to the brief summaries in other parts.

Topics

The most important part of your strategy document. This is where you list all the topics and issues which you intend to discuss, either in order of importance or split into functional area. In practice this part will usually be a very, very long table, where every row reflects a different issue or risk. The typical table will cover several important factors:

- **The issue**: Describe the topic which needs health authority feedback, in clear and unambiguous language.
- **Company position**: A statement of the position which the team will take on this issue. In other words, what is our 'answer' to the problem stated in the issue column?
- **Key points**: Covers the logical arguments which support our position, showing the reader why we think our answer is the right one.

- **Available information**: Describes the data which supports our arguments – so the studies or publications which have been performed or which are planned to be done. The description does not need to be highly detailed, but should be enough that the reader can understand and comment on the strategy.
- **Missing information**: What other data *could* support your proposal, but simply isn't available at the moment? Or may not be completed within the required time-frame?
- **Risks**: What risk does this topic pose to your development program? If you don't get an answer, what would it mean to your current status and your overall chances of success? This is a very useful way to prioritise questions for discussion.
- **Fall-back options**: Perhaps the health authority rejects your proposal, or your argument isn't strong enough to reach the briefing book stage. What do you do instead? And what will that mean for the project?
- **Briefing book question**: What question do you intend to ask? This will change many times before the briefing book is finally submitted, but it helps to get the first attempt in place now. Use clear language and focus on your goal.

Remember that the strategy document doesn't need to have 'complete' responses in place at this stage. You can happily add tentative hypotheses into the table or describe different options which will be taken under different circumstances. Then, as more data comes in, the strategy and rationales can be expanded upon or modified.

In addition to this, an issue is rarely confined to just one specialisation or field of study – the majority will require input from multiple experts. This is nothing new in drug development, which is a highly-collaborative area, but it can be helpful to explicitly mention the involved areas in the strategy document, including their required input. This saves a lot of time when assigning work later on.

The strategy document creates an 'issues focused' summary, it forces attention onto the potential problems in the development program rather than the program status. This is well-suited to preparing for an advice meeting (after all, why bother going to FDA if you don't have a problem to discuss with them?). It is also very useful when dealing with upper management, who can quickly gather an overview of the pressing topics and how they will be solved.

A strategy document is thus a central location for the issues which need health authority feedback, one which everyone can view, discuss, and ponder. It can be thought of as the physical manifestation of the team's thinking.

Step 5: Expanding into a detailed document

The initial draft of the strategy document will be fairly sparse, often with brief comments or 'to be determined' notes scattered around. The next step is to look into the topics in more detail – do the arguments make sense? Is there other data you could bring in? Are you describing everything correctly, or is it a cryptic word salad which no-one can understand?

As the team mulls over the issues and their response, the issue sections will become larger and more fully-formed. Eventually you will find that the guesses and hypotheses turn into clear statements supported by solid data from source documents. In other words, you now have the core of the eventual briefing book.

At this point you should begin asking for outside opinions – functional heads, management, etc. Every review round will bring new insights (sometimes valuable, sometimes just irritating) which may change your team position. Perhaps more importantly, this also makes upper management feel involved in the process and thus less likely to change things at the last minute.

Eventually all of the different stakeholders will have had their say and all of the comments will have been incorporated into the strategy document. At this point, congratulations! You have a finalised document, and can go start on the briefing book itself.

Does it have to be a document?

The ongoing expansion of the initial list naturally leads to the strategy document growing in length – indeed you will often see strategy documents reaching tens (even hundreds) of pages as more answers for each issue are brought in and described. This is not necessarily a bad thing (though it may not say great things about your overall risks).

However eventually a list of issues and risks to discuss may simply be too long for a typical document. Luckily there are different options:

- Excel is well-suited to organising information in tables, and makes it very easy to include new columns for responsibilities, etc. Filtering and automated marking of terms is also quite easy to do. General word-processing abilities are rather lacking, however, so the document will never be as easy to read as it could be (or at least, not without significant extra work).
- More structured formats such as relational databases or wiki-formats can also be used to map out the different issues and responses. This provides a lot of analytical power when working through issues, but adds a significant amount of complexity to the overall process.

- Alternatively, you can simply remain with a table in a (rather bulky) Word document. This has the advantage of being easy to use for all involved, though you may find that the program slows down and crashes more often under the strain.

Whichever of these options you choose, you should remember that the most important part of any strategy document is the ability to get everything in writing with a *minimum of effort*. The system should not be so slow or unwieldy that team members give up and store their information in a private location – this completely defeats the point of having a central location for knowledge.

CHAPTER 4: THE INITIAL APPLICATION

Once you have the strategy in place, it is time to request a scientific advice meeting with the relevant health authority.

The content of your request will be in line with the issues you identified in the earlier sections. Health authorities prefer fewer, high-quality meetings rather than multiple ones with single topics. Thus no matter your topics, you should try to group them together and provide multiple questions on the same product wherever possible.

In line with this, it's expected that you will have at least *tried* to answer your questions using the commonly available sources – guidance documents, online information, webcasts that may be streaming, etc. Ignoring this and asking a *really obvious* question will quickly lead to an outright rejection.

Application Content

The meeting request is a combination of 'general' information (which relates to your product and intended approach) and more specific sections (describing issues and the meeting you would like to have).

Health authorities provide guidance or template documents which are useful for putting together your application, though there is usually some flexibility in the structure allowed if you need it.

Both the initial application for a meeting and the briefing book will be submitted via the HA electronic gateway and will be linked into the existing CTD format (just as for any other submission).

General information

Your application should contain general information so reviewers can identify your project and gain a quick overview of what you are doing.

Thus you will need to state:

- The **identifying number** of your application. This may not yet be available as it is assigned by the HA following your first submission or interaction. But if you have one, include it.
- The **name** of your product. This can be a final marketing name or (more commonly) an internal company code for the molecule.
- **Specific information** on the molecule, such as the chemical name, the established/IUPAC name, and the structure if it is simple to provide (i.e. this is not required for biologicals).
- The **intended regulatory pathway**. This is more of an issue for the FDA, who like to know which regulatory pathway you intend to follow for the final submission (such as a 505(b)1, etc.). Nonetheless the other authorities will also appreciate this information.
- Intended **indications** for the drug. You should have an idea of what you are trying to treat with this drug, so this should be explained for the reviewers.
- **Study plans** which you intend to follow. This can include high-level overviews of intended clinical trials, paediatric studies, etc.

Meeting-specific information

In addition to the general project information, you need to declare how you wish to meet and how you intend to go about it.

This means your application should include details such as:

- **Type of meeting**. Each health authority has a set of different interactions which are possible, and they won't know what you need if you don't tell them. So you need to state if the meeting will be a Type B meeting, a BPD Type 4, etc.
- The **suggested meeting slots**. Possible date ranges for a meeting should be given, taking into account any absences of critical team members. (In theory, at least. In reality most companies will push for the meeting to be held ASAP and will require the team to cancel any holidays). You can also push for a specific time in the morning or afternoon. This is very important when international teleconferences across time zones are occurring.
- The **meeting format**. You can specify the format which you would prefer (e.g. for FDA you may ask for a videoconference or a written response rather than an in-person meeting). The HA can take this into account, or may simply make their own decision.
- **Timing** of the briefing book. Generally it's best to send the briefing book or meeting package along with the initial application, but if you are sending it later then you need to specify when they will receive the documents.
- A brief **summary of the purpose**. You should be able to quickly explain and summarise the issues which are occurring and the objective of the meeting. This should be short, as full details come in the later sections.
- A **list of participants**. You should provide the name and title of the people who will attend from your company, as well as any extra consultants, translators, etc. You should also list the expertise which you need from the regulator side – such as statistical experts, microbial validation experts, etc. It is possible to

specifically request certain staff members but this may prevent a meeting slot within a reasonable time. Note that the selection of experts from your side is a matter of strategy, as covered in the following sections.

Problem-specific information

The next section of the application should cover the specific problems and discussion to be covered. You do not need to include large amounts of detail here – this is just the application for a meeting after all – but you should provide enough information that the HA reviewer can understand the context of your questions.

The most important part of this is the list of questions which you intend to ask. These should be clear and focused, with a brief explanation to provide the background context for your request.

Keep in mind that it should *really* be brief! The main part of your argument will come in the briefing book and meeting package, and health authorities dislike being flooded with too much detail right away. If you are submitting the briefing book alongside the application then you can simply refer to that document for the full details.

The questions are the most important part of the application – after all, why go to this effort if not to get an answer? They get a correspondingly high amount of scrutiny from the health authority, and so blatant errors in this section will quickly lead to a rejection.

In general it is better to pool topics and ask several questions rather than splitting them over multiple meetings. However you need to be careful that you don't ask too many questions. The meeting time is usually fixed for an hour, and if the discussion is still on question 3 of 12 when the time is up then, well, too bad. Thus you will often be forced to accept written feedback on the majority of topics to provide discussion time for the most important.

Assessment by the regulatory authority

Once the application has been submitted it will go into the assessment queue of the relevant authority. There are different timelines depending on the type of meeting you have requested and (naturally) the health authority you are dealing with. We've summarised some of the more important timelines in Chapter 2 covering regulatory bodies in the major regions.

It is important to remember that the final decision on your request will be made by the health authority. Every application will be assessed, yes, but *they* will decide if a meeting will go ahead and what the format should be. It is very common to ask for, say, a face-to-face meeting with the FDA, only for them decide that your question is not really that important and thus deserves a Written Response Only (WRO).

Some meetings tend to be one format or another. From an FDA perspective the pre-IND and Type C meetings are often held as written responses, though they make an exception when proposing a new surrogate endpoint as these trigger a lot of discussion. Type B meetings are usually accepted and will often be face-to-face, though you will usually only be granted one meeting per development stage. It can also be difficult to get a meeting accepted if you are 'only' reworking a known API into a different dosage form.

These aren't hard rules. Drugs which have been granted special statuses such as 'breakthrough therapy' are granted additional meetings to help bring the product to market sooner. Similarly if your project is having multiple issues in different areas then you may be granted meetings for clinical, quality, etc.

All of this is useful to know but is completely outside your control. And so at some stage you will receive a response from the agency telling you what they have decided.

- If your meeting has been **denied**, then the health authority involved needs to tell you why. It is completely their decision, but denial needs to be based on a valid reason – i.e. having ugly formatting or filling in the application form incorrectly are not sufficient reasons. The two main reasons for a denial are *not including enough information* in the application/briefing book to discuss the issue, and feeling that *it is too early* in the development process to address the topics. You will also get a denial if your application is missing a lot of the required information which was mentioned above.

- As a better outcome, your meeting can be **granted**. In this situation you will get a notification of approval and a summary of the meeting format and timing. If it is a face-to-face or teleconference meeting then they will indicate when and where you will meet, how to attend, and provide a list of the staff members likely to attend. If you receive a written response then they will indicate the date when the written decision will be provided.

The next steps

Once you have received feedback from your respective health authority, then it is time to begin preparing the next steps. What those steps are relate to the authority feedback and the reason for their decision.

Your meeting was denied

This is an unfortunate situation and you can expect to have some sort of unhappy comments coming in from upper management. Leaving aside the implications for your job (or whoever eventually gets the blame), you should be thinking about *why* the application was denied. The reason for denial will inform your next steps.

If you have been denied because you **didn't provide expected information** then you can simply update the application/briefing book and resubmit. All timelines will reset, so you are basically starting again from the beginning, but with a much higher chance of acceptance. Hopefully you get it right this time – make sure you carefully read the relevant guidelines, use any templates which are available, and don't make the same mistakes twice.

If you were denied because you **didn't provide enough context** then you should go back into your application with a critical eye. Do your questions make sense given the information you provided? Is your background information enough for a non-expert to understand the problems? Ask people outside the project for feedback here. If you only submitted an application then you may prefer to include a full briefing book in the next attempt. If you did provide a full briefing book and they still cannot understand your problem then you may need to go back to the start and rethink your strategy.

Alternatively, you may have received a denial because it is **too early** in the development process. This covers questions which need certain datasets to answer fully, but where the data has not yet been created. So you could theoretically have a meeting, but there would be nothing to discuss and you would not be able to get a clear opinion from the health authority. If this occurs, then you need to go back and develop more data on this topic before resubmitting the application.

Or perhaps you were denied because you **didn't follow published guidance**. This occurs when the answer to your question is found within one of the many guidance documents which health authorities put out each year. The denial will politely refer you to a specific document or two, then suggest that you come back when you have a new question to ask. This is highly embarrassing and dangerous for your career, so always, always search online *before* you trigger a scientific advice meeting.

Your meeting was granted

The much better (and far more common) outcome is that your meeting was granted by the health authority. Feel free to have a little party or celebratory cup of coffee, after which it is time to get back to work. What you need to do will depend on the situation.

If you provided a meeting application **without a briefing book**, then it is time to start writing. Hopefully you have already started – the timelines for getting meeting packages in after acceptance can be very tight if you aren't prepared. In some cases you can go over the deadline, but more than a day or two is unlikely to be accepted. So grab the strategy document and move on to the briefing book.

If your application **included a briefing book**, then you have significantly less stress. Now is the time to begin preparing for the meeting – think of possible feedback or questions, decide on your official response to these, begin practising your answers, start drafting the presentation you need. Although most of the work will happen just before the meeting, it never hurts to start a bit early.

Finally, you may be **receiving written feedback** with no chance to discuss the topic. If so, then you basically have nothing to do until the response arrives. Yes, you will have to think up different possible reactions and decide on strategies for each. But most of the time will be spent reassuring upper management that the feedback will arrive and that they should not be worrying so much about things.

Chapter 5: Writing a Briefing Book

The briefing book (also known as a 'meeting package') is easily the most important part of the entire scientific advice meeting process. It sets the stage for your arguments, provides the data which you leverage to make your point, and is your best chance to influence the regulators towards your preferred approach. Thus you have to spend as much time as possible making it clear, focused, and simple to understand.

In addition it should always be *on time*. Health authority staff will hold multiple meetings to discuss your briefing book, particularly where it may impact policy or programs from other companies. They don't like rushing this sort of thing, and forcing them to do so (by, for example, submitting a late briefing book) will directly and unnecessarily antagonise them.

Best practice is to send the briefing book along with the initial request for a meeting. This is required for some types of meeting and optional for others, but it is generally best to send it at the start to provide as much information (and persuasion!) as possible right away. Passing the briefing book at a later stage can make sense if you are short on time and need to start the process immediately – but keep in mind that writing a *good* briefing book will take much longer than you expect.

Briefing books, just like meeting requests, are submitted via the electronic gateway and become separate files within the overall eCTD structure.

Timelines for briefing book creation

Every briefing book is different, and thus the time required to create them will also be slightly different. However as with strategy documents, it is possible to estimate general timelines for the various stages of the briefing book creation.

- **Creating from the strategy document**: The first step is to take the summarised information and planning from your strategy document and move this into the briefing book template. Use the document structure (questions, etc.) which you have already planned, and create placeholders for data tables and figures based on the agreed-upon logic. This process will take approximately two days.
- **Add required content**: Once the briefing book scaffold is in place, it is time for the contributors to fill in the details – the tables, the figures, the text arguing your points, everything. This can be the role of regulatory affairs, specialist writers, subject matter experts, or a combination of the three. It is important that the writing is clear, persuasive, and follows the aligned strategy. Time taken varies with the amount of content, but will generally require two to four weeks of work.
- **Review by the project team**: The project team are those experts who represent their individual functions, help make overall project decisions, but who may not necessarily be part of the briefing book preparation. Review by these experts should focus on *clarity and completeness*, based on their general if not detailed knowledge of the topic. This review stage will take up to one week, and implementing comments may require up to a week in case of major problems.
- **Review by functional management**: These are the managers of your experts, and they are (usually) experienced people with a long history in the field. They should review the degree to which the approach from the strategy document has been transferred into the briefing book. They *should not* try to rework the strategy to something new. As with the previous stage, this review

may take up to one week, implementing comments may range from a few days up to a week as well.

- **Review by upper management**: The outcome of scientific advice meetings can significantly change development programs, thus higher management will want to have a final say. This review should focus on the *clarity and persuasiveness* of the overarching strategy, ensuring that it has been accurately moved from the strategy document to briefing book. As before, review may take up to a week, implementing comments may be the same.

Thus the entire process from beginning (blank document template) until the end (a finalised briefing book) will take around 6-8 weeks. Complex topics or ones requiring a lot of background information will take longer, as will an initial health authority interaction – here we usually estimate an extra two weeks for writing. If a briefing book is being repeated (e.g. an EMA document reworked for the FDA) then timelines can be significantly shortened.

If you are responsible for organising the writing and review process, make sure everyone involved knows the timelines from the start. It helps to send out general calendar blockers to everyone involved ('review briefing book this week'), as well as specific meetings for comment resolution, etc. Whenever you send these out, make sure the strategy document is attached to the invitation (or otherwise easily accessible) – this helps prevent reviewers from trying to redefine the aligned strategy at this late stage.

It may also help to define specific sections for each reviewer using the commenting function of your word processor. This can seem a little heavy-handed, but it is important to avoid 'that reviewer'. You know who they are, the one that makes hundreds of pointless little comments throughout the entire briefing book. If you can't avoid them completely, then at least keep them contained.

What goes into a briefing book?

An important rule of thumb for briefing books is that they should contain enough to support the arguments, but no more. You don't need to include every last result in the document, merely the data which is directly relevant to your question. Anything else should be summarised and the results either moved to an appendix or left out completely.

Health authority decisions are science-driven and so they will expect to see data. Not just the data supporting your arguments (although this is obviously ideal) but the data which will help them to make a decision in the topic. In other words, don't hide results because they don't look wonderful, be open and discuss the implications.

This is particularly important when there is an obvious problem or a topic which is known to be challenging. For example, if you are aiming for a surrogate endpoint or a non-inferiority trial, you can expect the regulator to have many questions and comments on your approach. Approach these *openly* and talk about the pros and cons – this is far more likely to succeed than hoping the negative aspects are somehow overlooked.

The briefing book has several different sections, loosely grouped under general, meeting-specific and problem-specific information. Keep in mind that although this information should be *present*, you can choose where it goes in the document. This allows you to focus reader attention on the most important parts.

General information

There is a standard set of details which you should provide with every briefing book covering the product itself. You will notice that it is essentially the same as that provided for the initial meeting application.

These details are not particularly interesting except as background and so it is often worth moving them to an appendix to ensure focus remains on the issues and questions. Make sure you check with your local experts or procedure manager before doing this.

This general information includes:

- The **identifying number** of your application. This may not yet be available as it is assigned by the HA following your first submission or interaction. But if you have one, include it.
- The **name** of your product. This can be a final marketing name or (more commonly) an internal company code for the molecule.
- **Specific information** on the molecule, such as the chemical name, the established/IUPAC name, and the structure if it is simple enough to provide (i.e. this is not required for biologicals).
- The **intended regulatory pathway**. Do you have a new molecular entity, a biosimilar, a combination product, a device, etc. etc.
- Intended **indications, dosage form, route of administration and dosing regimen** for the drug. You should have an idea of what you are trying to treat with this drug and how you are going to go about it. This should be explained for the reviewers.
- **Study plans** which you intend to follow. This can include high-level overviews of intended clinical trials, paediatric studies, human factors studies for devices, etc.
- A **background section** for the product. This should provide a short history of the development program so

far, the current status, and a notification of any big changes which you are intending for the development program. You should also summarise any discussion which you've previously had with the regulatory agency.

Meeting-specific information

You will also need to provide some information relating to the meeting itself

- A brief **summary of the purpose**. You should be able to quickly explain and summarise the issues which are occurring and the objective of the meeting. This should be short, as full details come in the later sections.
- A **planned agenda**. This is a bit of a time-waster as the main topics will change several times before the meeting due to internal politics and preliminary feedback, but you should at least try. Provide agenda items based on your list of questions and estimate how long you will be talking about each one. Remember that the most important questions should go first.
- A **list of participants**. You should provide the name and title of the people who will attend from your company, as well as any extra consultants, translators, etc.

Problem-specific information

This is the main part of the briefing book and the section which should be most prominent when reading through. In theory it only needs to contain two things, but they are two large and important things:

- **Questions** for the health authority. These should be in line with your documented strategy and meeting application. They should be grouped by discipline (clinical, labelling, quality, etc.) and come with a piece of text explaining the context of the question.

- **Supporting data** for each question. What you need obviously varies depending on the question, but it should always be *summarised* material. In other words, the most important results should be presented in the briefing book, every last single detail of the study is not required.

For more detail on the best way to include all of this information, see the following section.

How to write your briefing book

Writing a briefing book is a specific skill which needs to be *practised*. You won't be perfect at it when starting out, this is why you have different review stages. Nor is every briefing book the same, the topics, content, and logical arguments will differ each time.

We go into a lot of detail on writing in the appendix (page 121). But there are a few basic pointers which are simple to keep in mind and useful in basically every situation.

- No waffling, no extended sentences. Write clearly, without extra words, and get your point across.

- Follow the 5/5/15 rule of thumb. Every page should have more than 5 paragraphs. Every paragraph should have less than 5 sentences. And every sentence should have less than 15 words. This forces you to write concisely and avoids the paragraph-long sentences which plague scientific writing.

- Avoid using 'and' to link different thoughts (just write two sentences instead).

- Other terms to avoid include 'and/or' (confusing meaning), 'moreover' (and other filler words), or nonspecific phrasing such as 'it' (if there is the chance of readers getting confused).

- Try to use active voice whenever possible ("we consider" rather than "it was considered") – this is easier to understand and in line with FDA recommendations.

Simple and clear writing is even more important for briefing books which will require translation (for example when meeting the Chinese NMPA or Japanese PMDA). Make sure it is clear for the translator and double-check that they are clear on complex terminology or argumentation.

The following sections will cover some specific factors to keep in mind when writing the different parts of the briefing book.

Front matter

Every briefing book will have a front page listing the drug under discussion, associated development codes, the authoring company, and the date. Your company should have a defined 'style' for documents – one with specific formatting and logos, etc. Make sure you use the correct template here, because some authorities have specific requirements for text size, spacing, etc.

If you don't have a set style or template within the company, health authorities such as EMA and FDA have template documents which can be downloaded and used as a basis.

The early pages should also include a list of abbreviations. In general, if you use an abbreviation within the text then it should be listed here. This is not necessary for very common terms (such as 'DNA') or those only used once (in which case it should be explained in the text). Standard abbreviations used in a non-standard way should be listed here as well.

Introduction

The introduction should mention your drug (and what it will treat) and then directly lead into the intended goal of the scientific advice meeting.

There is often the temptation to include extensive details of development history or mechanisms of action, etc. The reviewer usually doesn't care about this as much as you think. Given their lack of importance compared to the main topic, it's usually better to shift it to a later section or appendix.

New information

If you are having a second interaction with the health authority (or third, etc.), then you may find yourself repeating a lot of information from previous briefing books. A nice way to avoid this is to simply refer to the content of the previous document, and then describe new or changed information in this section. Make sure that you provide enough background for the reader to understand what has changed.

List of questions

This section will cover all of the questions which you intend to ask the health authority, and as such can be considered the most important part of the entire document.

Questions are grouped by discipline (e.g. CMC questions, followed by clinical questions). When grouping questions, make sure that the most important disciplines and topics are at the beginning. This focuses attention on the topic and gives you a higher chance of a full response. Generally the order of questions should have been defined during strategy document creation. If the order is not defined, you will need to discuss with the project team.

Each group of discipline-related questions should include a short introduction section giving important background information.

Try not to repeat information which comes in later parts. Instead focus on providing context for the more detailed discussion to come

There are many ways to present the questions and supporting data. Our preferred approach is to begin with a short introduction and background to the topic, followed by the question to the health authority.

There will always be some degree of background information and context which needs to be provided. You need to include enough that the issue can be understood, but not so much that they drown in details. A novel or unexpected topic will require more background than a standard question on a well-established theme.

Wording your question

Properly wording the question is surprisingly difficult and will tend to trigger a lot of discussion. You should have a first approximation from the strategy document, but now is the time to really get into the details and make sure it is perfect.

The most important goal is to be clear, precise, and simple. What do you need to know? Focus your question on that. Don't try to hint at your meaning, don't ask general or vague questions in the general area. Both of these risk off-topic or incorrect feedback. Make sure you **directly ask** about potential problems rather than vaguely requesting feedback on your approach (which, oh, by the way, coincidentally includes a solution to a problem we're seeing, would that maybe be ok? Possibly?).

You have some leeway in how long the 'question' is, and it's perfectly OK to use a few sentences to get your point across. In other words, don't create horrifically a long and torturous 'sentence' to ask a question. It's also fine to write a statement of your position and then ask "Does the agency agree?"

But sometimes what you are asking is simply too complex for one 'question'. It is possible to use sub-questions (so Question 1a, then 1b, etc.) but this tends to make a muddle of the following data and company position. It can be simpler to use separate questions, each leading towards the next one. Always keep the end reader (the health authority reviewer) in mind – what can you write that will ensure their full understanding?

Questions can be split into two main groups, open-ended questions and closed:

- **Closed questions** are looking for a straightforward 'yes' or 'no' answer. They are the sign of a company with a good idea of what they want to do, and they want a clear sign from the health authority if it is acceptable or not. These questions generally have a lot of data and results supporting the company position. You will see them more on advanced projects where much of the development work has been performed.
- **Open-ended questions** are more exploratory, seeking to understand the health authority opinions. You have an idea, but you aren't really sure and want to get feedback. Do they like it or hate it, and if so, why? These questions tend to be more common in early development, where many issues are still being worked out.

The choice of question is an important one as it heavily influences the information you receive. Closed questions will give you a clear answer, but perhaps not the one you are looking for. Open-ended questions will receive a lot more thought and background information, but may not provide an answer.

You can maximise the chances of getting good feedback by including context for your proposal, discussion of other options, and comparing their positive and negative aspects. Feel free to nudge them towards your preferred choice, but make it clear you've thought about several options.

Questions should be numbered and defined as 'headings' in the document – this automatically places them into the Table of Contents where they can easily be found by the reader.

The company position

The question should be followed immediately by the company position, which is essentially what you believe is the correct answer. This is your chance to direct the reviewer towards your thinking, something which is a good idea even in the most open of open-ended questions.

The company position should begin with an executive summary – a simple statement of your position. This is then followed by a slightly more detailed rationale – the reader wants to know 'why' you have this particular opinion, and it is your job to answer them. Importantly, you should provide this using logical arguments and persuasion, don't just summarise data and hope it works out.

As with all writing you want to be concise, using clear and positive language with a focus on facts. Any information given should relate to the question being asked. It should be provided in enough detail that someone reading *just* the question and company position will be able to understand your argument.

But avoid going into too much detail – try to limit this section to high-level or summary data which clearly supports your position. If you have highly detailed information (clinical study reports, etc.) then this can be included in a separate appendix section. This keeps the main document tidy and easy to read.

One helpful tip is to include hyperlinks to this appendix data only at the *end of the section*. Hyperlinks within the text are temptingly easy to click on, which can pull the reader away from your carefully-constructed argument. By placing all links at the end, you force them to read through your entire masterpiece before they can leap away.

Feel free to use subheadings, bullet points, etc. to improve the structure and readability of your text whenever necessary. There is nothing worse than pages full of words with nothing to break up the monotony. Remember that reviewers can get bored as well!

At the end of the company position, restate your main point as an overall conclusion. The standard rule of persuasion applies here – **say what you are going to say, say it, then say what you said**.

Background information

The background information section provides context for the document *as a whole*. It is the information which isn't linked to any one question, but rather your overall project. You should think carefully about what should go in here, or even if it is required at all.

Very long or detailed background sections should be split up such that the overview or summary remains in the main briefing book, while full details can go into an appendix. Remember that you don't need to repeat this every time you hold a scientific advice meeting, subsequent documents can just refer back to previous briefing books.

Product background

This will contain a selection of fairly standard information about the drug you are developing. This includes the chemical name, the structure, the mechanism of action, the proposed indications, proposed dosages, route of administration and dosing regimen (assuming you have gotten that far, of course). All of this information can be placed in a neat table, provided in detail once, and then simply referred to every subsequent time.

Regulatory background

This is the section to briefly describe the current regulatory status of your project, as well as listing any scientific advice meetings which have been held in the past. It is less useful these days as most countries use electronic dossiers and thus can easily find previous submissions within the eCTD. As such, it's usually better to write a short summary or even skip entirely.

Quality / CMC background

The CMC section is usually prepared by the CMC team due to the specialised knowledge required. It should be tailored to match the questions which you are asking in the briefing book. The amount of background information needed depends on the context – are you asking a single, fairly understood CMC question or a lot of highly technical ones? Don't explain the basic things (especially to knowledgeable health authorities like FDA and EMA) and don't put vital information here either (that goes in the list of questions). But it should set the stage for your questions.

Initial/early meetings generally focus on clinical trial design rather than CMC topics and so can get away with a paragraph or two of very general background. For example, say your process has been transferred to large scale, but don't bother giving any more than this.

Later meetings tend to bring the commercial manufacturing process into scope, which means more CMC information is required. You don't need to go into details, for example you can say the drug is made in a XXX litre bioreactor with mammalian cells rather than giving exact processes and cell lines. This of course doesn't apply if the details are directly relevant to a question you are asking – in this case, add more information!

When asking about comparability or similarity exercises (tech transfers and biosimilar development, for example) then make sure you have a *summary* of tests / criteria / justification at the start before you go into the details. Nothing is quite as terrifying to an FDA reviewer as 20 pages of information on randomly assorted analytical methods and their acceptance criteria – eyes will glaze over and reviewers will fall asleep. Avoid this, summarise!

Non-clinical background information
As with the CMC information, this should be tailored to the questions you are asking. Include things such as pharmacology and toxicology, outcomes of studies which you have performed, and estimated dosing regimens based on that.

In general this is specialised enough that the non-clinical expert on the team should be writing it directly.

Clinical background information
This can be a very large section depending on the stage of development you have reached. Include a summary of clinical plans, results obtained to date, ongoing, completed and upcoming studies which may be related to your questions (for example, trials in other indications). Remember that it is primarily a summary, so provide only the relevant information in an easy-to-understand manner. If you have a lot of detail then this should go into an addendum to the briefing book.

Over time, as you have multiple meetings with the health authority, you will find that the summary data changes as well. This is normal, after all, you are developing a new drug. But make sure you highlight the updates so that the reader can easily see what has changed – this is a simple way to avoid off-topic questions from the reviewer.

List of references

This should provide full details of every reference (academic or other) which you have referred to in the briefing book. Check authority guidance for the exact format which they look for, but in the worst case you simply need to ensure that anyone can find the reference with a quick online search.

It is no longer a requirement that you attach copies of every single reference paper. However it is very helpful to attach copies of the critical publications so that the reviewer can jump back and forth directly. However you do it, make sure it is easy to see what is provided and what is not.

Appendices

Appendices provide supplementary information which supports your questions and associated company position. Despite the temptation, this is not the place to throw every bit of data you have lying around. It should be *focused*.

Appendices can be provided as separate PDFs linked into the eCTD or merged into the main briefing book document. Be sensible about how you do this, particularly as very large documents tend to be unstable on older and slower computers.

There are many different things which could go into an appendix section. Here are a few ideas:

- More detailed summaries of results, data tables, figures, process descriptions, statistical analyses, and more
- CMC study reports such as validation reports, comparability, and so on
- Investigators brochures
- Development plans
- Draft of a study protocol or synopsis
- Draft CTD modules

- Copies of regulatory guidance from other countries (mostly when talking to smaller authorities, make sure you don't provide copies of their own guidance – this is considered highly insulting!)
- Drafts of prescribing information
- Prescribing information of approved competitors or reference products
- Minutes from previous health authority meetings

CHAPTER 6: PREPARING FOR THE MEETING

The briefing book is away and the health authority reviewers are going over it with a fine-toothed comb. Which means you have a little bit of time to spare. So it's time to get ready for the meeting!

Who will attend?

This is a very important topic for the success of your discussion, and it requires a bit of thinking beforehand. (Actually, you should have thought of this before sending in the briefing package, but we'll assume that you are simply reading ahead in the book).

There are three main groups of people that should be at any scientific advice meeting: the subject matter experts, regulatory affairs, and decision-maker representatives. We'll cover all three in the following sections.

Subject matter experts

These are by far the most important people that you can have at any scientific advice meeting. They are the ones with deep knowledge of the problem you are dealing with, the possible solutions, and the arguments for one solution or another.

Ideally you want people with both experience and flexibility, as they will have to deal with many unexpected discussion topics during the meeting. Naturally they also need to have the social skills to persuasively interact with others in a business environment. This is, unfortunately, not always the case. Pick based on ability, not just seniority.

The experts present should cover all of the questions which you are planning to ask. If you are asking specifics of product analytical testing and statistics for a clinical trial, then there should be people who understand these topics. It's fine to have multiple experts to each function, but don't bring too many. Crowding the meeting won't help with the discussion.

Regulatory affairs

Regulatory affairs are the next set of people that should be present at the meeting, for fairly obvious reasons. They (hopefully) have experience dealing with the health authority in question and are familiar with the overall process.

Their job is to act as a bridge between the authority representatives and your team. They will host the meeting from your side, providing introductory discussions, naming people attending, and handing over to the correct expert for discussion. They are also responsible for taking notes, summarising discussion points as you go along, and keeping everyone within the time limit. This last one is often the most difficult.

It's worth having at least two regulatory affairs experts in the meeting. This allows one to focus on guiding the discussion while the other takes notes. It also makes meeting preparation in the months beforehand far easier when several people can share the load.

As with subject matter experts, you want flexible and experienced people to be present in the meeting. Leadership skills are also important – there will be a lot of pressure as the big day draws near and regulatory will need to help guide the team through it.

Decision-maker representatives

Decision-makers are those people who have relatively high positions within the company and so have the authority to decide on a course of action for the project. In other words, if the FDA requests another arm to your clinical trial, they are the ones who can approve the (many) extra millions needed to do so. Their specific position depends on the size of the company – it might be a department head in a multinational or a CEO in a small start-up. The important thing is that they are able to make decisions.

Now, and this is important, these people **should not be at the scientific advice meeting**. Their ability to make binding decisions is dangerous as their off-hand comments can bind you into taking a certain approach.

What do we mean here? The ideal advice meeting involves a lot of discussion between experts on difficult topics, eventually leading to a middle-path which fulfils requirements on both sides. Ideally, you will get a strong statement from the authority that *if* you were to take this approach, they *would* approve it. But you *should not promise* to take this approach during the meeting.

Why? Circumstances change. You may discover other alternatives after the meeting, new data may come up, the clinical endpoints may change. More likely, someone on a Steering Committee will disagree with the costs and demand you do something different. Whatever the cause, there are many occasions where the post-meeting approach doesn't exactly line up with the in-meeting compromise. And if you had previously agreed to this compromise, perhaps through the approval of a decision-maker, then you're in trouble.

Thus decision-makers should not be present in the meeting. But a **representative** from the decision-makers should be there. Their role is to keep focus on the company strategy and provide information flow between the meeting participants and upper management. But they can't agree to anything, and so the company is safe.

Issue cards

By this stage you have spent a lot of time thinking about your topic. You've assessed data, you've looked at the literature, you have a good idea of the scientific basis of your arguments. You should also have a good idea of the *gaps* in your arguments, the areas which don't quite hold up so well.

If you don't know what these are, it's time to go back and work on the strategy document. Come back to this point when you've worked them out.

It's OK to have gaps. No scientific development program is perfectly solid, because no program has fifty years to figure out every last complication. You are in a hurry to get to market and have a limited budget. And that means you will have gaps in your knowledge.

What is important is the nature of the gaps and your ability to compensate for them. A trace, but unidentified impurity in your drug which is below well-accepted levels according to literature? Not too bad. A potentially severe interaction with the immune system which you haven't really looked into? Definitely a problem (and something you should fix, honestly).

Whatever they may be, you should be aware of gaps and you should try to mitigate their impact in the meeting itself. One way to do this is with **issue cards**.

Creating issue cards

An issue card (or challenge card, or trigger card, there are lots of names) is actually pretty simple. It is a simple statement of a problem (in this case your knowledge gap), written on a bit of paper. You then use this as a launchpad to discuss the problem, your mitigation actions, and how to persuade others that this is acceptable. Once cleaned up, this creates an 'official' response to that topic which the team members can follow in the meeting.

Let's cover this in a bit more detail.

- *Identify your gaps*: As mentioned, you should know what these are. Work through them based on importance. You (or the team) should create a simple one-sentence description of the gap – enough detail that everyone understands it, simple enough that it can be skimmed in a rush.

- *Write each gap on a card*: Card, piece of paper, whiteboard, it doesn't really matter. The important thing is that everyone present can see and discuss it.

- *Discuss it*: Sit down with the people who know about this gap. This could be a couple of the experts or half the team, and often for major cross-functional gaps you will find that it becomes a very big meeting indeed. Whoever they are, make sure all attending are involved in the discussion. Your aim is to identify the arguments that help you cover this gap. Feel free to challenge solutions which are suggested, you want the argument to be as solid as possible.

- *Create a story*: Pull the arguments together into a simple narrative that covers the gap. Using the gap from before as an example, you could think along the lines of "We observed an impurity but couldn't identify it. However we could quantify it as 0.X%. This is below threshold Y based on literature A, B, C and regarded as safe based on FDA guidance D. As such we consider it acceptable."

- *Write this on a fresh card*: The old card is likely covered in scribbles and mind maps. So take the moment and create a new card with the knowledge gap and your narrative response. This is now the aligned and official team response if the gap is challenged. The relevant team members should try to memorise it (as you can't bring a cheat sheet into the meeting room).

- *Repeat for the other gaps*: Unless you are in the miraculous situation of only having one gap. You're either incredibly lucky, or you missed something.

At this point you're probably wondering why you should bother. Isn't most of this part of the strategy document from before? And yes, often these problems are covered in the strategy document. But it's very, very rare that anyone *remembers* what was in the document, let alone bring the arguments to mind in a stressful meeting environment.

Issue cards help by creating a simple, easy-to-follow response to each problem. They can be shared amongst the team and quickly memorised, allowing you to prepare for a range of challenges.

Which is the next thing you need to practice.

Challenge sessions

Discussing options and memorising arguments is good to start with, but it doesn't really prepare you for an actual scientific advice meeting. A meeting is high-stakes. Everyone is under pressure, you need to think fast and be ready when things go completely off your carefully-prepared rails.

Luckily you can still practice for these situations prior to the meeting. Different organisations use different terms – rehearsals, challenge sessions, hostile Q&A, mock advice meetings or grilling sessions. They come in a variety of flavours depending on how prepared the team members are. But the overall aim is the same – to somehow replicate the circumstances which the team will experience during the scientific advice meeting.

Light rehearsal

The easiest option is a simple rehearsal of the meeting. It usually focuses on the prepared sections (your presentation, initial remarks, and so on) rather than the discussion and questions. A typical rehearsal will involve the following steps for the meeting attendees.

- *The team gathers:* The entire team comes together in a central location (if it will be a physical meeting) or online (if partly via video-conference). This requires the subject-matter experts and regulatory affairs, sometimes with management representatives in attendance as well.

- *Meeting lead starts the meeting*: This is normally regulatory affairs or someone else with experience in advice meetings. They describe the events which are occurring – we all dial in, introduction to the call, FDA states their names, we state ours, etc. This doesn't need to be completely detailed, but enough to give an idea of what will happen.

- *Presenters present*: Assuming you have some sort of presentation to start out the meeting, you would then run through these slides. This is a chance to see what works, what needs fixing, and where the presenters need encouragement.

- *Meeting lead summarises between sections*: This doesn't have to be much, just say 'and now we have discussion on topic X'. It helps get everyone into the mindset that it is multiple topics to be discussed, not just one.

- *Meeting close-out*: The meeting lead summarises the topics which have been presented, thanks everyone for their time, and closes the meeting. Everyone sighs with relief, and then discusses what could be improved.

This is a very gentle way of rehearsing. It focuses on the fixed sections and allows everyone to get a vague idea of how the meeting will run. There are opportunities for improvement in a low stress environment. This approach is most suitable for a new or inexperienced team, or immediately after creating the final slide deck prior to the meeting.

Mock advice meetings

Rehearsals are nice but they won't prepare you for a real meeting. This approach takes it a few steps further by adding representative health authority members who will ask questions.

As before, the team should run through the introduction, opening the meeting, presenting, and so on. This is a fixed section and the more times you practice it the easier it will be.

However in addition you should select several people *from outside the team* to play the role of the health authority. They should have a good understanding of the topic at hand, but they should not have been involved in the meeting preparation. You can think of them as knowledgeable outsiders. Ideally they have also seen an advice meeting or two and so know how the process goes. They should have time beforehand to study the briefing book as well.

Their role is then to ask questions during the meeting. Good questions, bad questions, tricky ones, whatever takes their fancy. Their aim is to pick away at the presentation and find holes in the arguments. How well can your team members explain the information they present? Handle unplanned questions on different topics? Are their answers understandable for non-experts, but still detailed enough?

Its worth noting that it is fine to pressure the meeting team. But it should be done in a professional way. Real scientific advice meetings don't descend into screaming (usually, at least) and your outsiders should hold themselves to the same standard. Making people cry doesn't help team morale.

Once the mock advice meeting is over, the outsiders should give their feedback. They should point out good points and bad points. If team members did well, highlight this. If answers were poor, rushed, or badly explained, point this out as well. The team can then take this feedback and improve for the next round.

You should aim to have 2-3 mock advice meetings during the preparation process. A single meeting is not enough to properly incorporate feedback. Four or more is generally overkill unless the meeting is very important (Advisory Committee and so on) or you expect a lot of difficult discussion.

Hostile Q&A

Rehearsals and mock advice meetings are normally enough to carry you through most situations. Your team members also have plenty of other things to do and so can't spend the entire time practising.

But sometimes you have a very important meeting coming up. One where you expect there to be a lot of discussion, a lot of pointed questions, and where your company position is likely to be heavily challenged. Advisory committees are a nice example of a meeting where this may occur.

In these situations you need to practice answering the very difficult questions. As before, you will need outside experts to play the role of the health authority members. However this time you encourage them to be nasty. Ask difficult questions, press for answers, reject explanations directly if they are not acceptable, hammer on certain points as though you are making a political statement. You may need to find additional volunteers, often including upper management, to ensure the appropriate level of pressure.

Facing this kind of questioning can be very difficult, even if you know it isn't 'real'. There are a few tricks which you should keep in mind.

Keep calm
Often the most difficult bit. But the best way to deal with hard questioning is to remain as calm and collected as possible. Breathing *out* (not in) and counting to three or four can help your thoughts settle before you respond.

Work out the type of question
Hostile questions generally fall into one of four types.

- *Negative preconditions*: These start by referring to a negative example (e.g. this failed in the past) and then link it to the current situation. Focus on the current situation and current issue, and make it clear why it is not intrinsically linked to the past issue.

- *Negative comments about you that are presented as a question*: Generally rare in scientific advice meetings, but sometimes emotions flare up. Just ignore it and move on to the scientific topics.

- *Normal question, hostile tone*: A normal question which is asked in an attacking manner. Ignore the approach and treat it as though they asked politely.

- *Provocative language*: Using language which makes things sound more dramatic or horrible than they really are. Respond using your own words and don't echo theirs.

Connect to the core of the question
Questions can be tricky and nasty, but there is always a core topic which is being asked. Your job is to find that issue and acknowledge that.

It's OK to make statements such as 'I understand you are asking about X' – this gives you time to think about your response. You can always ask for clarification of the question if it's unclear – 'what aspect of Y concerns you?' By doing so the questioner is forced to rephrase, often in a more calm way.

Once you have the core issue, focus your reply, calmly and politely, on that. Imagine that the question was asked in a neutral manner and respond in kind. Remember that the only way to 'win' in an advice meeting is through scientific reasoning. Focus on that, not the emotions, in order to succeed.

Redirect the question
This is difficult to do well but can be very useful in some situations. When faced with a hostile question, provide a brief answer but then immediately bridge to a persuasive argument which you want to use.

Putting all of this into practice
How does all of this work in practice? Let's suppose you are hit with a nasty question: "Can you deny that your drug will kill children through it's terrible safety profile?"

- This is not pleasant to receive in a meeting, even a practice one, so our first step is to remain calm. Big exhale, count to three, etc.

- Next we work out what this question is really asking. It is highly provocative and emotional, but the key words are 'safety profile' and 'children' – the core issue is one about paediatric safety.

- We ask for clarification to calm things down, ("what part of the paediatric safety profile are you concerned about?") Note the neutral tone and that we are not volunteering information yet.

- The questioner then has to provide more detail, e.g. "there were several serious adverse events in clinical trial X, which imply…"

- Now you have enough details to acknowledge the question again ("I understand you are concerned about…") and provide your response to the issue ("in-depth investigation revealed that these were due to…").

Is it easy? No. That's why we practice.

CHAPTER 7: PRELIMINARY FEEDBACK AND FINAL PREPARATION

The last few days before the scientific advice meeting will be a whirlwind of stress and last-minute preparation. Preliminary comments come in, the final strategy needs to be aligned, you have to prepare and practice your presentation, you need to be clear on how the difficult questions will be answered...

Sounds like a lot? It is! You should expect to be working fairly continuously for the 5-6 days prior to the meeting, regardless of whether a weekend occurs or not.

Preliminary feedback from the health authority

The health authority will provide preliminary feedback shortly before the meeting occurs. For the FDA, you can expect to see a PDF with their current thinking about 5 days prior to the planned meeting date.

It's important to remember that this feedback is *really* preliminary. Although it shows the thinking of the health authority involved, this can change significantly as a result of discussion during the meeting itself. Which, after all, is the whole point of having the meeting. So take anything written as a *starting point* for your discussion.

The only situations in which preliminary feedback should be considered final are a) if you decide that the meeting is now unnecessary and call the whole thing off, or b) decide that *this question* has been answered sufficiently and thus decide to ignore it in favour of other, more important topics in the meeting.

And don't forget – this is *not* the time to bring in lots of new information to argue with the health authority position (or worse, try to ask new questions on the topic). That time is past. You can certainly sneak some things in that weren't available in the briefing book. But anything which can be considered 'substantial new information' will be rejected or ignored.

Understanding the preliminary feedback

The feedback you get is based on a number of factors. The briefing book reviewers will have their own opinions, they will then meet internally to discuss your proposal and their preferred response. Additional experts will be called in if the topic is one with strategic or policy implications – this is particularly common when your project is ground-breaking or highly controversial.

Regardless of who writes it, the feedback you get should be read very carefully and with an eye for the underlying arguments. Health authorities are notorious for giving ambiguous responses or ones which subtly suggest a course of action which they cannot openly endorse. Do not take the words at face value, *think* about them.

After you've really, *really* thought about it, then you can classify the response into one of several distinct possibilities.

Straightforward agreement

This is the simplest and most useful response that you can get – the health authority agrees with your proposed approach. In this case simply remove this question from the meeting agenda and use the time to focus on more controversial topics.

Dossier review issue

The health authority agrees that you do have an interesting question, but does not believe you will have enough data to meaningfully discuss it prior to dossier submission. As such, this is considered to be a 'review issue', one which they will look at in the course of assessing your New Drug Application and so on. This is clear but frustrating feedback as it essentially wastes your time – you now have no guidance for solving your current problem and leaving everything until dossier review is a very risky approach.

If you do get a response like this, try to identify the *most important piece of feedback* which you need. It may not be necessary to have a complete answer for your entire question, but a clear statement on one part could help get your development program moving again. Once you identify this critical feedback, put all of your focus on getting an answer. This applies even if the rest of the question needs to be left unanswered.

Clear reference to guidance

The health authority takes your request and respond with a reference to a guidance document or two which they have previously published on the issue. If they are being particularly clear they will even include a quote from the guidance document showing exactly what they have already said on this topic.

This situation can be extremely embarrassing. It usually indicates that you didn't look into the existing regulatory documents before sending off the meeting request and the authority is now gently rubbing your nose in it. You will have to think fast to avoid problems with management and other internal stakeholders, who will understandably want to know why you missed it.

This of course relates to keeping your job. From a project perspective it is perfect feedback – just implement the advice quoted in the guidance document, whatever it may be. On the other hand, if the authority has misunderstood the question and the guidance is not applicable, then you should prepare to argue the point in the meeting.

Ambiguous commentary

Sometime you will receive feedback which doesn't appear to make sense. It restates a position which you already know, it discusses topics which don't seem relevant, it refers to apparently random data within your briefing book.

Although it seems to be confusing, this is actually a valuable opportunity for your project. It often indicates that the health authority *wants* to give you a certain answer, but is held back by confidentiality or policy reasons. In other words, there *is an answer there*, you just need to figure out what it is.

In these situations you need to get as many experts involved as possible, don't just rely on the inner circle. Try to think of different meanings they could be assigning. Look at the exact wording, go into the topics which they discuss. These potential meanings can then be used as topics during the meeting itself, where you can look forward to a challenging and sometimes frustrating discussion while trying to determine what exactly they want to tell you.

Straightforward disagreement

The health authority comes back and says, quite clearly, that they disagree with your position. They will then suggest an alternative approach which is in line with their thinking and expectations.

How you deal with this is very dependent on the suggested approach. It's *usually* best to follow the advice if they are asking for something minor, simple to implement, or in line with general regulatory expectations. In these cases you have a clear indication of their expectations for the final submission. Pushing back will not achieve anything and you can expect to receive the same comments during dossier review. Pushback also annoys your counterparts in the health authority, which makes all subsequent discussions much more difficult.

Things get more interesting when their suggestion is wildly different to your proposal. Look at your arguments again – was this a reasonable company position? Do you still believe it makes sense? Do you have credible arguments as to why your approach is better than the suggested one? If so, then it is worth trying to push back on their proposal.

Note that this requires that you have excellent arguments and data on your side. You will have trouble in the best circumstances as the authority has already judged your data through the briefing book and found it lacking. Nonetheless it is possible to win the argument in the scientific advice meeting if you are lucky and persuasive. So why not give it a try?

Finalising your strategy

The previous steps involve many, many meetings. You'll be discussing strategy within the program team and with upper management… and to be honest pretty much anyone who walks by the office will try to put their own opinion in as well. Try to keep things as minimal as possible, though don't be surprised when it doesn't work.

Nonetheless you should have a couple of goals during these initial discussions.

- **Decide on your overall strategy**: In line with the previous section, what are you going to do? Push back? Agree? Accept feedback you don't like? Try to change one critical factor?

- **Determine the priority of the questions**: A briefing book normally has a number of questions for the health authority. Some are more important than others, and this importance often changes once the preliminary feedback is in. You only have an hour, so now is the time to prioritise the discussion points. What will you focus on, what will you leave out?

- **Decide on the response you will provide**: Once you have a strategy, you can decide how to implement it. Which data will be highlighted or expanded upon to support your point? Can you bring more persuasive arguments than those in the briefing book? Should you pivot completely and try something different? Stress-test your ideas – what questions could the authority ask? What if they refuse to believe certain pieces of data? How can you compensate?

- **Decide who will be there**: The participant list was already shared as part of the briefing book. But you might realise that different experts are needed – some topics are gone, some are new. In this case you need to get the experts on-boarded into the topics and ready to discuss.

- **Get management alignment**: This is dependent on the culture in your company – some are very top-down, some allow program teams to make their own decisions. But for critical topics or feedback it is important to get upper management informed (and preferably in favour of) your overall strategy.

- **Make a presentation**: We'll cover this in the next section.

Always remember as you go through that there is a limited amount of time in the scientific advice meeting. Typical FDA meetings go for one hour, and at least 5 minutes of that will be taken up with the initial welcome/introduction.

Thus you need to be ruthless with prioritisation of topics and time. There is nothing worse than reaching the end of a meeting and realising that you will not be able to discuss one or two critical topics.

The company presentation

It is not a requirement to give a presentation. And indeed authorities such as FDA often say that it's better not to, given that they have received all the relevant information in the briefing book. But despite this almost all applicants use a slide presentation to start off the meeting.

Why? It's a good way to get the company line directly in front of the authority panel and allows you to guide the flow of the meeting. But it is only effective if the presentation is well made – and this is what this section will cover.

By this stage you should have worked out your overall strategy and the specifics of your responses for each question you want to discuss. For each question create some slides to describe your position and highlight relevant supporting data.

The slides should be short and crisp. Do not flood the slide with data tables or huge amounts of tiny text. It should focus on the main message without allowing the reader to be distracted by strange colour combinations or extraneous detail.

The executive summary

The executive summary is often the most difficult slide to create. It should summarise your key arguments, storyline, and supporting evidence. The intention is to allow the reader to quickly understand the most important information.

Most presentations in the pharma industry will use the standard situation-complication-resolution framework (i.e. what is the current situation, what is the problem which has arisen, what do we recommend you do about it). This is not always the best option for a scientific advice meeting, particularly when there are multiple questions that need to be discussed.

If there is a major theme of your questions, then it is usually better to have an overall executive summary which focuses on this but which is intended for *pre-reads only*. The health authority members will receive it beforehand, and you can use this to guide their expectations. In the meeting itself you can skip the executive summary and go straight for your issue-related slides.

If you have a vastly different set of questions, then trying to make an executive summary will take much of your precious preparation time and probably only end in confusion. Focus on preparing the individual slide sections as well as you can, making sure the arguments are solid and in line with your planning.

Individual slides

The major part of your presentation will be the slides covering the individual topics. These are very, very important – you need to get your arguments across in a manner which is both convincing and not horrifically boring (regulators are rarely convinced when they are falling asleep, after all).

Slides are built from three main components, the title (which tells you the key insight of the slide), the subheading (which summarises where that insight is coming from) and the body or content (which provides the data and visualisation).

A handy rule of thumb is that the title should not say anything that's not supported by the slide body, and the slide body should have nothing that is irrelevant to the title.

- The **title** should be an 'action title', one which describes the key message of the slide. "Impact of clinical trial size on power" is not an action title, but "Enrolment of 500 patients allows accurate determination of efficacy" is. Clearly spell out what you are trying to say, don't allow the reader to interpret it themselves. They may get it wrong.

- The **subheading** is not a necessity but can make explanations easier. It should give a short description of the data source used to make the claim in the title. Think of things like "Anion chromatography results from validation batches" or "Average incidence rate in 2023, per thousand".

- The **slide body** should present the relevant information in the simplest possible way. It is tempting to include all the data which you have thought up to support your argument. Resist the temptation! Too much information will only dilute your main point and thus weaken your argument. Use graphical visualisation wherever possible versus tables of numbers, and make sure all text is short and to the point.

Make sure you follow a consistent format and colour scheme. Most organisations will have some sort of official slide set-up, make sure you ask for it. If this doesn't exist, pick a few simple colours and stick to it.

Generally the less you have on a slide the more memorable and persuasive it is. Large blocks of text or crowded data tables in small font will guarantee that your audience tunes out. To avoid this, use graphs, figures, and visualisations wherever you can. Cut down text wherever possible to simple, clear statements. And don't be scared to use icons or labels to emphasise key points in the information you present.

Your overall aim is to make each slide *insightful* (providing information which is not obvious or widely known), *actionable* (contains something which can be used to make a decision on your topic), and *relevant* (actually relates to the topic you are discussing).

This is not an easy process. You will likely have several rounds of slide preparation, discussion, and then modification. Probably under significant time pressure and with stressed colleagues. You can avoid this by reading up on effective slide creation and communication in advance (and there are many good books out there, see the further reading section on page 185). But realistically you probably won't have done this, so just dive in, iterate your way forwards, and remember that it doesn't have to be perfect.

The slide 'storyline'

Most topics which require a scientific advice meeting are too complex to be explained on a single slide. The 'storyline' is the flow of arguments and evidence across one slide to another, all focused on the one topic.

There are a few key things to remember here:

Start with the conclusion

This feels strange to people with a scientific background, we are used to writing scientific publications where the reader is led through our results before reaching the discussion. But this doesn't work in high-pressure situations with limited time to assess and think about the data. Instead you need to lay out the **conclusion first**.

This is often referred to as the 'Pyramid Principle' and it is core to consultant communication (and thus by extension very handy in scientific advice meetings). State the main argument as your first slide ("the new manufacturing process is suitable for commercial use") and then the relevant insights come in the following slides ("the quality attributes are statistically identical", "the site has been successfully audited by the FDA", etc.). The data which supports these are then either summarised in the individual slide bodies or (if highly detailed) moved to the back-up slides.

Slide titles should tell the story

The audience should be able to take your slide deck, read nothing but the slide titles, and *still* be able to understand the both main conclusions and how you reached it. In other words the slide titles should act like a story of their own. This is why the idea of 'action titles' mentioned above is so important.

Use section dividers

This seems pretty obvious, but a surprising number of slides move from one topic to another with barely a break in between. Use mostly-blank dividing slides to mark a clear cut from one question to another. It also makes it clear to everyone in the meeting that you are finished with one topic and are ready to move to the next.

Polishing

Once you've got the main slides, storyline, and so on in place, it doesn't hurt to do a quick polish of the presentation. Is this a vital factor which will sink your scientific advice meeting? Well, no. But a polished presentation gives the health authority an impression of competence, and this 'halo effect' leads them to trust your arguments more.

Also from a purely career-oriented viewpoint, your slides will be reviewed by higher management, who see many well-polished presentations each week. It's better to be remembered as a good example than a bad one.

What are some typical mistakes to watch out for?

First is the challenge of **inconsistent formatting**. The entire presentation should follow the same formatting: this means using a consistent template, font and font size, colour scheme (for charts and text as well), as well as consistent use / location of your company logo.

Next you need to watch for **missing details**. These are the little things like chart titles, labelled axes, legends on charts, footnotes to describe data sources and assumptions, page numbers and so on. It is a small thing, but people will notice.

You also want to check for **unprofessional information**. No stupid pictures, memes, weird text choices, cartoons, and so on. Nothing reduces your chances of success like a wildly-out-of-place image in a serious document.

Don't forget to check the **title page**. Is the date correct? The project code and drug name? The presenters? This is particularly common if you are reusing a previous slide deck, perhaps even from another health authority meeting. Internal reviewers also tend to ignore this slide when reading the deck, so be sure to check it yourself.

Reviewing and rewriting

It would be nice to think that your work is done at this point, and that it is now just a matter of rehearsing and preparing for the meeting. Well... no. Because now is the typical point for all of your different management levels to start reviewing the presentation.

You can avoid the main annoyances by getting management alignment on the strategy beforehand (as mentioned in the previous section). There is nothing worse than completing a fantastic presentation, then changing the entire thing because someone important fundamentally disagrees with your approach. Align things up front!

If you do hit the worst case (major changes to strategy are requested), then you should be ready for a lot of discussions and meetings. Try to find out what their underlying concerns are, look at alternative options within the team and with management. Sometimes you will find a compromise, sometimes you are forced to change. It sucks. We've all been there.

Hopefully you will be luckier, with comments that focus on data presentation and arguments. Good ideas to improve your persuasive approach are always welcome, incorporate them whenever you can.

Comments which focus on low-level topics such as layout or wording can be ignored or incorporated as you want. Management review should not be focused on these topics so you can push back by claiming this is the style of the presenters. However this (obviously) will not make you popular with the manager suggesting changes. Choose whatever makes the most sense for your project and your career.

Finalise

Everyone is finally happy with the presentation? Fantastic! Now it's time for the final clean-up before sending it out.

Read through one more time, very, very carefully. You are checking for anything which lowers overall professionalism, which includes things such as:

- Removing any comments which were added, including 'resolved' comments which are sometimes hidden by the software.

- Removing in-line written commentary, highlighting, text saying "delete me when we have the data", etc.

- Changing the document author name to that of your company

- Fixing spelling and grammatical errors. You will still find typos on the meeting day, but try to avoid it as much as possible.

- Alignment of all elements on the slides both within the slide (e.g. text lines up the the heading) and across slides (e.g. heading is in the same place each time).

- If you have included hyperlinks or interactive elements, are they working properly?

- Do the slide numbers match the presentation order?

- Is the document reasonably sized, or do you need to compress the size of multimedia elements?

Gone though it? Everything looks good? Then it is time to send it to the health authority as part of your pre-meeting package.

Response to the health authority

There is a general expectation that you will provide your presentation and overall intentions to the health authority before the meeting. Rules vary, but it is at least 2 days prior to ensure they have enough time to review and discuss.

The response package should include the finalised presentation which you have created. It should also describe the questions which you intend to discuss – if the meeting will focus on clinical questions and CMC topics will be dropped, then say this. This allows the health authority to set their expectations and sometimes reshuffle their expert attendance as well.

Rehearsals

Once the slides have been created, rehearse the presentation. You should aim to rehearse in front of the entire project team and at least a few of the most-involved management levels. This is not easy, particularly as you want the audience to act like a hostile reviewer. Get them to ask nasty questions and challenge your opinions. You are particularly looking for ways to improve presentation flow, adaptability, and persuasion.

Presentation flow covers the way in which you work through the slides. Is your timing good? Not too fast and not too slow? Are you engaging the audience and speaking from memory rather than reading off the slides? Do you pause for emphasis rather than rushing through? Are you speaking clearly, in a varying tone, emphasising key words and phrases?

Adaptability covers your ability to react when things don't work perfectly. Can you handle questions? Are you able to pick the right person from the team to cover the question? Are you calm, rather than freaking out? What happens when the audience member stops you and says 'I don't believe you'? Yes, it happens, and no, it isn't nice.

Remember that you are an expert on this topic, you and your team have been working on it for (at least) several months by now. You should have enough of an overview to direct the questioner in the right direction, towards the relevant data, towards the correct team member. In particular the issue cards which you created earlier (page 80) will help you answer correctly. Keep calm and carry on.

Once you've rehearsed, gather feedback from the people present. Are there weak spots or areas which don't flow? Can you improve your presentation to fix these? Did you completely screw up and need to avoid ever making one particular comment ever again? Take everything on board, think about how you can improve, and improve.

And then rehearse it again.

CHAPTER 8: HOLDING THE ADVICE MEETING

The big day has arrived and you are about to enter the scientific advice meeting.

As we've mentioned before, the point of these meetings is to have a *collaborative* discussion on a topic which is causing problems. It's not an argument, it is not a chance for you to lecture them on your brilliant idea. Instead you are there to present challenges, let them ask clarifying questions, and seek the health authority's opinion on the best path forward.

This is not to say that you are just wandering in and letting them tell you everything. There is an expectation that you have already thought about the problem and the best approach. But your company needs to have an open mind and a willingness to discuss the solutions.

So how do you actually do this? The meeting day can be broken up into three main parts – pre-meeting preparation, the scientific advice meeting itself, and post-meeting debrief.

Pre-meeting preparation

Let's assume that you have already worked out your strategy based on the previous few sections (and if not… now is a bit too late). Thus the day of the meeting should be spent ensuring that everyone knows the plan and that everyone is ready to perform.

From a logistics perspective, you should hold a pre-meeting with all the people that will be involved in the advice meeting. This will either be in the hotel room (if you're travelling in person) or in the office/virtual (if you are having a teleconference).

Preparing for an in-person meeting

In-person meetings are held at the office of the health authority in question. This adds another layer of complexity for those who are attending, including the logistics of getting there.

Hopefully you have though about this issue before this point, and are not currently sitting on the other side of the world wondering about visa requirements. Any in-person meeting requires travel to the health authority location (you go to them, they don't come to you), which means you need to think about:

- Flights and other transport to the meeting location.

- Entry visas if you aren't a citizen of the country in question.

- Hotel rooms for at least the night before the meeting (because you don't want to be rushing to the airport two hours before it is meant to start).

- Suitable clothing – generally everyone from the company should be wearing a suit, even if the health authority attendees don't always bother.

You will also need to think about your business requirements. Do you have remote access to email, chat, and relevant files? There is often a question or two that comes up just beforehand, and being able to look up the relevant data directly is incredibly important.

Assuming all this has worked out, your team should travel together to the health authority office location. Do not discuss the details of your request in the waiting area, in the nearby cafe, in the toilet, or anywhere nearby at all. You never know who may overhear, and health authority employees pass on gossip just like the rest of us.

You will generally have to wait until just prior to the meeting, at which point you will be escorted to the meeting room.

Preparing for the teleconference

These are getting more and more common, particularly after COVID times.

Although virtual, it's far better for team communication if everyone involved is in a single location. Reserve a conference room in your building, drag everyone into the office for the discussion. Open a chat channel for all attendees (but keep in mind that speakers will be too busy following the discussion to read messages). The designated room leader should keep an eye on the chat, bringing critical information to those who need it.

Use the pre-meeting to check all online connections, particularly if you have multiple people in a conference room who are dialling in simultaneously. A normally-stable wireless connection is often overwhelmed when everyone is using videoconferencing bandwidth, so stress-test it beforehand.

Make sure everyone checks their audio/visual connections – microphones and headsets if they are isolated and central speakers if together. Cameras also need to be working properly – most of the big agencies now expect all speakers to be visible to help the discussion.

If you are dialling in from a remote, non-office location, then make sure you *find somewhere quiet* to call from. The middle of an open-plan office is not quiet, your kitchen table surrounded by small children is not free of distractions. Aim to minimise all of these factors and stay on mute whenever possible to avoid background noise. Also be careful to use a microphone and headset, not only for better audio but also to avoid echoes coming through the system.

The Meeting Itself

The big moment has arrived! Everyone is ready, you are waiting outside the conference room or about to dial in, and nerves are starting to spike. Or rather, they will be when you reach this point, because we assume you are reading the book well beforehand rather than five minutes prior to the meeting.

Regardless of when you opened this book, it is a good moment to cover key communication principles and the general flow/structure of the meeting.

Key communication principles

No matter whether you are speaking in person or calling over the phone, there are a few basic principles which you should always (always!) keep in mind. Many advice meetings have been wasted through bad communication, make sure it doesn't happen to you.

- **Listen to understand**: You are here to get advice from the health authority (or at the very least, to hear their viewpoint). Make sure you listen to what they are saying, let them finish their sentences and ask for clarification where needed. Many solutions are only hinted at by HA members, you will need to tease the answers out yourself.

- **Focus**: Pay attention and be mentally present – don't drift off into other topics or spend the time thinking of what you will say next (this last one is a constant danger, particularly when you are new and nervous).

- **Polite and respectful attitude**: *How* you communicate has a huge impact on success. Focus on presenting a calm, constructive attitude during the meeting – you are here to solve problems, not argue. Similarly you should avoid dismissive language or point-blank rejection of their comments. Body language should be engaged and interested, hold eye contact with the speaker and lean in when they are talking.

- **Be succinct**: Don't waffle on. Make your point clearly, without extra words, and then stop. Don't volunteer extra information which isn't needed, you will inevitably pull the conversation off-course.

- **Be smart**: As the name implies, it is a *scientific* advice meeting. You will achieve your aims by demonstrating that your position is well-grounded in scientific facts and logical arguments. Make sure this is clear when you present and discuss.

- **Focus on gaining the required information**: It is very easy to get pulled away from the main topic and into discussion of minor or irrelevant details. Avoid this! Constantly ask yourself if you are getting the information which *you need* for your program. If not, feel free to (respectfully) note this and pull the conversation back on track.

Typical roles within the meeting

There is a general flow which most meetings tend to follow, similarly there is a standard set of roles which you will see repeated across meetings. The main roles which you will see are as follows:

The meeting chair (authority)

The chair of the meeting will inevitably be someone from the health authority. They will do the usual formal bits and pieces (date, time, organisations present) and then lead into an introduction round of the authority experts who are present. This will include names and titles, giving you a clear indication of the main counterparts for your issues.

Attendees (authority)

There will be many, many different people attending from the health authority side. Most of them will be silent the entire meeting. However, just as in your case, they are there to step in when their expertise is needed (or simply to keep track of what is going on). Just ignore them, focus on the people who are speaking.

The overall lead (applicant)

This person acts as the host of the meeting from your side. They introduce your attendees, co-ordinate the conversation and responses, and summarise discussion as topics are closed out. This is generally a member of the regulatory affairs team, though you can use anyone who is quick-thinking and well-versed on the topics being discussed.

Speakers (applicant)

Speakers are subject matter experts on a particular field which is relevant to the discussion. They should be assigned to a particular question or topic in advance, and ready to step in with the company viewpoint (not their own!) whenever necessary. Try to ensure that Speakers only enter the conversation when the overall lead gives the OK. Failing to control this rapidly leads to a chaotic discussion with very little value to your team.

Presenter (applicant)

This is simply a speaker / expert who has been assigned a topic to present in the meeting. As they will be in focus during that topic, they should be experienced and prepared for the discussion to come.

Room leader (applicant)

The room leader isn't a necessity but can be helpful in larger virtual meetings where the team is distributed. Their job is to ensure that everyone is engaged and paying attention, sending subtle reminders via chat or text if needed. They don't direct the flow of conversation but will often mention important info from background or chat discussions to the overall lead where needed.

Note-taker (applicant)

One or two people should be designated as note-takers. This is not to say that the others won't be taking notes (they should!), but rather acts to ensure that the entire meeting is covered. Note-takers should ideally be fairly free of presenting duties, as this will obviously impact their ability to make summaries.

Note-takers are important in any meeting, but they are vital in scientific advice meetings because recording is almost always *not allowed*. The health authorities try to preserve the idea of open discussion, which suffers when everyone involved is watching exactly what they say. Thus there are no recordings, and the only permanent or official record of the discussion is the meeting minutes, which are, naturally, highly summarised. Don't rely on this, take notes!

Silent listeners (applicant)

The silent listener is one who is not on the official list of attendees, but who nonetheless sits in the tele- or video-conference room to listen in on the discussion. Without taking part, obviously. Silent listeners can range from upper management wanting first-hand impressions through to new members of the regulatory department getting some experience.

Keep in mind that most health authorities *do not approve* of silent listeners and often explicitly request that *only* those listed in the agenda take part. It is up to you (and your company, obviously) if you follow this or not. Just make sure nothing is written down.

Flow and structure of the meeting

Although each meeting is different, there is a typical flow which is followed throughout.

Introduction round

The chair of the meeting will summarise the date, time, organisations present; then lead into an introduction round of the authority experts who are present. After this it is time for the overall lead from your company to do the same, doing a round of all present (or dialled in to the teleconference), stating names and titles.

Note that in recent times it has become acceptable to simply refer to the pre-meeting information, which includes all of these names anyway, and have each speaker introduce themselves when they begin to talk. This works quite well and gets you into the topics a few minutes faster than is otherwise possible.

At this point the overall lead will hand over to the first presenter.

Presenter topics

The presenter should introduce themselves with their name and role, turning the camera on if it is a video-conference. They then present the assigned topic, using the final slide deck to facilitate discussion where needed.

Generally you don't need to spend time inside the meeting showing presentation slides. All of the important information should be in your briefing book, and so you should not be going over it in repetitive detail. However it can be helpful to create a *short* summary to restate the main topic. This should have been sent to the FDA a few days prior to the meeting and will allow you to reinforce your position, it will also help guide the discussion if everyone gets off track.

Again, it should be a short presentation. You will almost never get an extension of your meeting time, particularly if the time has been wasted in walking the audience through slides. Skip the presenting and focus on the important bit – discussion!

If the health authority members have questions, the presenter is the first one to answer. They should provide an initial response to the question, then hand over to other experts as needed. Use eye contact, chat, or just ask directly to ensure that they are ready to step in.

After each topic has concluded, you should have one person summarise the discussion to this point and the overall conclusions. This is useful for several reasons. It lets you verify that all present have the same understanding of the opinions (even if they are completely different between health authority and your company). It gives the minute-takers a chance to catch up with the discussion. And it provides a quick mental break before moving on to the next topic. You will also find that the summary will trigger additional comments or explanations from the HA staff, which in turn gives you more feedback.

Overall summary and close-out
Once you have worked your way through all of the topics, hopefully with valuable feedback for all of them, it is time to close out the meeting. The overall lead should summarise the topics which have been discussed, the main points to take away, and any action items which may need to be performed.

This summary is important as it allows you one last chance to get feedback from the health authority. If they disagree with your summary, they will say so. It also helps anchor the topics in everyone's minds before the post-meeting letdown pulls it away.

Responding to questions
The way in which you respond to questions has a big impact on the overall success of your meeting. Defensive answers, arguments, and so on will simply make your life more difficult. As mentioned before, you want to have an open, science-based

dialogue which is intended to persuade them to your point of view.

Answering an authority question generally follows a simple 4-step process:

- Thank the speaker for their question (not really required, but it's polite)

- Repeat/restate the question to ensure that you understand what they are asking. Allow them time to correct you in case you have it wrong.

- If you have already provided information in the briefing book or other documents which is relevant, tell them where it is so that they can follow along. This is often a simple matter of asking everyone to turn to Back-up Slide X and working from there.

- Provide your answer to the question, aiming to be as comprehensive as needed (but not more). If you need more information or insights from the rest of the team, hand the question over to the relevant expert.

Cancelling or rescheduling

Although there is a lot of planning and preparation that goes into a scientific advice meeting, things do change and you may find the agreed-upon timeline doesn't suit anymore.

Fortunately it is still possible to cancel or reschedule the meeting, although you do need to have a good reason. Whether a reason is 'sufficient' is entirely up to the health authority in question – you can ask for a delay, but they have the final say on the matter.

Some of the 'allowed' reasons for rescheduling or cancelling include:

- A slight delay in preparing the meeting package, such that it will be available in sufficient quality but perhaps a few days late. Here you need to request the delay and hope that they accept it.
- The health authority realises during review that you need more data to hold a proper discussion, but is aware that you can gather this fairly quickly. The meeting will be delayed based on agreement between the two parties to allow this information to be prepared for the discussion.
- You have gone all out and provided the authority with too much information, to the extent that they cannot review it all within the formal timelines. In this case the meeting will be delayed by the health authority to give themselves required thinking time.
- The company sends some additional questions to the health authority after the meeting package is submitted. In this case there will be a delay (pushed by the authority to give themselves enough time) or you will be told to submit another meeting request with everything. You should try your hardest to avoid being in this situation.
- The authority decides that they need more people than were initially planned, because your issues are actually trickier than expected. So they push the meeting back to get all of the new experts involved as well.
- The briefing book came in very late, or was written in a terrible and impossible-to-understand manner. The authority will then cancel the meeting and tell you to come back when you are more competent.
- Finally, the applicant may decide that the preliminary responses have answered their questions fully, and thus the meeting is no longer necessary and can be cancelled. The authority may disagree and force you to have the meeting in any case, particularly if they want to push a discussion on a certain area.

CHAPTER 9: POST-MEETING TASKS
Stakeholder management

Immediately after the meeting finishes (or as soon as practically possible) you should pull together the entire team that were in the scientific advice meeting. Consolidate notes, go over the discussion, decide what information you have received and how that relates to the questions you have asked.

It's important to do this *right away* because memories fade quickly, particularly in high-stress situations. Often the key answer to your problem will be a couple of seemingly-insignificant comments from a random expert on the panel, and you need to avoid this being lost.

Once this has been done, you will need to start managing the different stakeholders involved, particularly upper management. Either the regulatory lead or the program manager will draft an official team summary email, this will go out to various interested parties to stop them asking incessant and irritating questions. For a day or so at least.

You should use this time to pull all of the notes together and come up with a high-level understanding of what has occurred. In particular you want to be able to answer the following:

- Where did the authority agree with your proposal, where did they disagree?

- What does this mean for your project?

- Will you need to do extra studies and create more data?

- How long will this take? What will it cost?

- Was there some wriggle room in their comments, or was it a clear decision?

- Do you need to ask for more money / time, or can it be covered with the planned budget?

- Do you need to provide something to the authority in the immediate future?

These are all important because they impact your project strategy and future planning. Negative feedback can force costly delays, and in the worst case may lead to the project being cancelled or de-prioritised internally. These decisions are normally above your pay-grade, and this is why everyone will be asking for additional information as soon as possible.

Answer openly with the knowledge and assessment that you have available. Remember that the *final* decision is only made with the written feedback, but management will want to make *a* decision now. Support where necessary, but make it clear that it is still only preliminary information.

Providing meeting minutes

Someone within the company should have been taking notes throughout the meeting – in fact, ideally several people will have been doing this. You should consolidate these notes with the recollection of people involved either directly after the meeting or (at latest) on the day after.

Once you have consolidated notes they should be provided to the health authority via your procedure manager. Obviously they should be cleaned up and understandable – no-one appreciates getting ten pages of random scrawl.

Company meeting minutes should be sent to the health authority within 1-2 days of the meeting itself. The authority is under no requirement to take your minutes into account, and indeed can choose to ignore them completely. However it is worth taking the opportunity to slant final minutes slightly in favour of your company position. This doesn't always work, and is perhaps a bit underhanded, but it can be surprisingly effective.

Final feedback

The FDA aims to get meeting minutes back to the applicant within 30 days of the advice meeting. Keep in mind that this is a *target* and not a legally binding goal. If you are dealing with a particularly challenging topic, one which is politically controversial or which would set a wide-ranging precedent, then you can expect them to take longer. There have been in situations where we waited for months after a meeting while the FDA discussed the implications with their policy group.

This can be very difficult for the regulatory experts, particularly when a decision needs to be made quickly (or worse, when upper management starts nosing into things). It is acceptable to follow-up with the procedure manager once the normal deadline is past, and you can request more information on when they will be finished. Sometimes they won't say, sometimes you will get a date. Sometimes you can get unofficial feedback on a very specific question by directly asking the procedure manager, this can give you information you need before the full meeting minutes are available. Risky, but often worthwhile.

As mentioned above, it is polite to provide the company meeting minutes to the health authority. They are within their rights to completely ignore whatever you gave them and write their own. And whatever *they* write is the official version, not yours. In practice they will normally incorporate your comments into the minutes (often word-by-word), but you will see major differences when a topic has been controversial and heavily argued.

The official minutes are not a transcript, they are a summary of the discussion. It will point out agreement or disagreement, and further items for action or discussion. It will also be based on the preliminary feedback which was given to you prior to the meeting (depending on the discussion, of course). You may not get much in the way of detail, particularly on 'boring' questions, so make sure you take your own notes!

Sometimes the minutes will include additional information or explanation that wasn't provided during the meeting itself. This may be due to a lack of time, or perhaps because internal discussion was required. This information will be explicitly marked as 'extra' that was not part of the main meeting. So if you read something and cannot for the life of you remember when it was discussed, it's probably a post-meeting addition.

What if we disagree with their feedback?

Health authority feedback from a scientific advice meeting is *advice*. It is not a legally-binding requirement, it is not a set of laws carved in stone. It is advice which you are welcome to take or to ignore, the decision is up to you.

In general it is better to follow any advice which is given. The health authority has more experience on these topics than you, they see similar problems from other applicants. They also have a clear idea of *what they want* in the regulatory submission, which is obviously the fastest track to approval.

The meeting minutes are part of the dossier and the advice will be clearly present for the later reviewers to see. It will be obvious that you have ignored previous advice and this can affect your chances of approval. So a general rule – if they ask for something, particularly minor things such as additional assays or a slightly larger clinical trial, then just do it. Save yourself the hassle.

But sometimes you will disagree. The authority may not understand your specific problem, you may be asking something which is completely outside the normal ranges. Or you may not want to spend the money that would be required. In all these cases you can ignore their advice, though this obviously comes with additional risk.

The risk here starts with additional questions during the review process, often very pointed ones asking for the information you have skipped. In really severe cases you may have your submission rejected or receive a complete response letter. This is, obviously, bad for your project and bad for your ongoing employment.

Risks involved with rejecting advice should be minimised wherever possible. This means you need solid scientific reasons for not doing whatever they asked for. It needs to be better than whatever you included in the briefing book (as that obviously wasn't enough) – and may include additional experiments, further analyses, literature reviews, etc. Always use scientific arguments, never base your (written) arguments on financial grounds, i.e. "we couldn't afford to do this trial". They don't care about your cash-flow problems.

Complicated submissions with problems that are outside health authority experience will often require further advice meetings.

What if we disagree on the content?
It is entirely possible that you disagree with the health authority regarding the contents of the minutes. Generally if you don't like their feedback (i.e. they said your approach was wrong) then there is not much you can do – either change your approach or ignore the feedback. However if you disagree on the *summary of the discussion*, then it is possible to request changes.

This is not for minor mistakes, rather major discrepancies between your recollection and the FDA. When it happens, raise the point with the procedure manager first. They will have a look and provide a preliminary assessment. If you still disagree, then you will need to write a document describing the specific points which are wrong and then file this into the dossier. The FDA will then consider this and make a final ruling as to whether the minutes should be changed, or should stand as they are.

What if we think of a new issue?

During the course of briefing book preparation and holding the advice meeting you may think of additional issues to discuss. These may be directly related to the ones in your list of questions, or they may be new ones based on feedback. In either case, you would like to bring these up for more discussion with the health authority.

Well... tough. You will need to start the whole process again. Write a new briefing book, apply for a new meeting, go through the entire waiting period again, etc. etc. Is this a massive pain? Yes it is. And incredibly frustrating when you think of something really important. But this is a general requirement from all major health authorities, mostly to stop applicants from spamming them with never-ending lists of questions. So you will need to deal with the problem internally or begin the process again.

APPENDIX 1: WRITING SKILLS FOR SCIENTIFIC DOCUMENTS

In this section we will cover the basics of writing for a scientific and regulatory audience. This is not just useful for briefing books, you will find these skills to be relevant throughout your career in pharma.

An introduction to writing

At some stage you will inevitably be called upon to write something. This may be a report of the findings you've made in the course of an investigation, it may be a briefing book, it may be a memo to management to keep them aware of an important topic.

Regardless of *why* you are doing it, a well-written report is vital for getting your main points across clearly and unambiguously. By contrast, badly-written reports leave your readers scratching their heads and wondering what you were trying to say.

Unfortunately the majority of scientific or technical documents are badly written. This is not due to their technical complexity or level of jargon (although this doesn't help), but rather because most scientists can't write for *external* audiences. They understand their own work exceptionally well, and thus rarely think about what the *reader* doesn't know, and thus needs to have explained. This leads to logic gaps, missing explanations, or sections where a pile of information is thrown down without any thought of clarification.

Writing a good document relies on several factors, all of which need to be present for maximum clarity. These are:

- A coherent message or conclusion
- A logical structure

- A consistent storyline
- A clear writing style
- A consistent document style.

We'll look at these in more detail in the following sections.

Creating coherence in your message

Any document will bring together a score of little details and minor facts, collecting these facts into a larger whole. But simply listing facts is useless, you need to order them in a way which makes sense to support your final recommendation. This summation and analysis is vital for the end users of the report – the information and conclusions given will help them to make their own decisions. This is, after all, the reason you are being paid to write it.

Any document should be built around an overall message or recommendation. This can be almost anything, of course, depending on the information you have and what the data shows you. The important thing is that this central message be determined *before* you begin to write.

Why? Because you will structure the entire document to direct readers towards your central message – hopefully in such a way that they agree with you. Readers of badly-structured documents will need to draw their own conclusions from the information which you present, these will often be completely wrong. Arguments amongst readers about the *implications* of your conclusions are an essential part of any important document, arguments about how those conclusions were drawn are not.

To summarise: you must know the key message first before building the report around it.

Creating a persuasive storyline

The overall document and self-contained subsections will have a 'storyline', a story which the report tells the reader in order to persuade them that the central message is, in fact, the right one.

This can seem strange to beginners ("I'm writing a professional report, not a fairy tale!"). But you will find that effectively getting a point across to busy readers requires a flow of logic and information. In other words, a storyline.

There are two ways to construct a storyline, these are based on logical arguments or groupings of items.

- A **logical argument** says that X is true, which is supported by data from Y, which in turn implies that Z is true. It is best used when you are extrapolating from a sequential set of information.

- An **item grouping** says that statements X and Y are true, thus the conclusion is Z. It is best used when you have a number of parallel reasons for making a statement of claim.

Storylines exist at both the report and the chapter level – your sections will have their own mini-storylines which will combine into one greater one. Importantly, you can mix methods within the report – one chapter may use an argument, another may be a grouping set-up.

Storyline via logical arguments

Logical arguments are the easiest to understand, the entire report provides a chain of logic which leads through to the final conclusion. This usually works as follows:

- A statement or section covering the current situation begins the chain of logic. This is the 'what is happening now' part.
- A number of sections follow, each of which provide more information that narrows down the potential options or possibilities. These feed into one another, such that one set of conclusions helps set the stage for the next. These are the equivalent of saying 'we know [X], and based on [Y] we can conclude [Z]'.
- A concluding section then shows that, based on the previous logical links, the main message and conclusion is the correct one. This is the 'therefore we should do...' message.

Logical arguments are a good way to guide the reader through your thinking about the issue. It is a more directed way of pushing peoples' opinions to align with yours, taking them through the evidence one step at a time. This makes it a useful method when you are delivering an unwanted message or a conclusion which the reader is likely to disagree with. The chain of logic can soften an unwanted message and show that this approach is the only reasonable one.

The disadvantage of this approach is that the reader needs to keep all of the preliminary information in mind before getting to the 'therefore' of the conclusion. It also requires you to have tightly-fitting arguments. If the audience can poke holes in *any* part of your logic then the entire conclusion will fall apart.

Logical arguments work at any level of your structure. When the main storyline of the report is based on a logical argument, you should be careful to focus on the 'new' information you are providing. Don't simply restate the preliminary information that everybody is aware of, use the chain of reasoning to bring your new data to the fore and support *your* final conclusion.

Storyline via grouping items

A grouping approach does not follow this chain of reasoning, instead placing a number of related thoughts or facts together. For example, you could group problems or challenges required to reach a single goal, reasons to take one approach over another, or evidence for a statement. As a rule of thumb, if you can make a dot-point list of a number of statements under a single heading, then you have a good set of candidates for grouping.

Grouped approaches ignore the step-by-step reasoning of a logical argument and instead persuade the reader with large amounts of information. It is particularly handy when you need to encourage discussion of a number of conclusions or recommendations. The list-like format helps to emphasise that you are drawing conclusions from multiple angles, while getting your point across in a clear, straight-forward way.

This also has the advantage that arguments are provided in 'parallel', if one of the arguments is rejected by the reader then the others may still be comprehensive enough to persuade them. The downside is that you abandon the hand-holding approach of the logical argument, which means that the report may be too abrupt and forceful for some audiences.

Structuring the document

Once you know the key messages and the storyline, it is time to work out how they will be brought across. And the first step to doing so is working out the structure of your document.

Start with the solution first

One trap which many ex-scientists fall into is that of writing a report as though it were a research paper. In other words, you spend a while talking about the observations which were made, then bury the recommendations in the 'conclusion' section at the end. This is perfectly fine in the academic world, where readers are generally interested in *how* you came to your final conclusions. In business, however, managers and executives are time-poor and want to see the conclusions *first*, then decide if they care enough to look into the details.

This is not to say that you should not present all of the relevant data, but rather that you should never *start off* with the data. Many executives will not care about your reasoning, many will have made up their minds on the issue already and simply need your report to support their intended actions. Thus they will often read the first few pages, say 'great', and move on.

This is sometimes irritating, yes, but it is also how things are at the upper level – an executive is paid to focus on the larger picture rather than the fine details.

Sorting out your thoughts via pyramid structures

The structure of a report goes a long way towards helping people understand it. Essentially, a structure helps you to group ideas and information to support summaries and conclusions, which in turn come together to support your final statement.

You can think of this as making a large pyramid, where a number of facts make the bottom layer, each of which is grouped to support a summarising statement directly above it. These are then further grouped to suggest more abstract conclusions or forecasts (e.g. extrapolations from the pooled data), which is then further grouped to support the main points at the peak of the pyramid.

The pyramid should consist of grouped ideas and facts rather than groups of topics. Summary blocks, for example, should say what the summary is (e.g. '10 of 23 drugs were approved last year') rather than having a nondescript label (e.g. 'drug approvals'). This forces you to make your conclusions explicit while putting the pyramid together and in turn forces you to be clear in your thinking. The end result is usually a pyramid of boxes, each containing a sentence or two to help nail down the meaning of your idea.

Similarly each of your extrapolations should be able to link into other ideas. Do not just summarise the summaries, show how they link together and thus how you can make the next statement on the pyramid. This helps to pull the entire structure together and avoids having loose statements flopping around and cluttering everything up.

All of this seems like a lot of work, and indeed it is. But it is a vital first step in putting a report together. Once you have you pyramid of ideas together then actually writing the report is much easier.

An example pyramid of facts

Facts	Summaries	Extrapolations	Main point
Incidence rate of Cancer X was 150 per million last year	Cancer X diagnosis has increased by 50% in the past year over historical average	Cancer X cases are expected to grow further in the future	Oncology drugs targeting Cancer X are an interesting development target for our company
Historical incidence rate was 100 per million		Oncology drugs have greater potential sales than other areas	
Expected lifespan is currently estimated at 85	The population of elderly people is expected to increase in the future	than other areas	
Historical expected lifespan was shown to be 70			
Sales of oncology drugs are currently at 10 bUSD per annum	Oncology drugs have greater sales than other therapeutics		
Sales of other drugs per sector average 5 bUSD p.a.			

You now know just what facts and ideas need to be included and how they relate to each other. Each level of the pyramid can be thought of a steadily-decreasing heading level – your extrapolations may be top-level headings, your summaries may be lower, your facts may be present in tables. But by linking everything together as in the pyramid, your overall argument and structure will transfer across. Even better, as each section is relatively self-contained, you can begin writing at any point you wish. Remember to add a few pieces to link sections and tie the main conclusions or message together, and there you are. A report. If you need to rearrange the sections for improved emphasis then it is a simple matter of cutting and pasting.

Developing your structure

The nice thing about the pyramid-based approach to structure is that you can start it at many different levels. In general, however, the most useful ways to plan out a structure will be top-down and bottom-up.

Top-down structure

Top-down is most useful when you already know what you will be saying or what you will be talking about. It can also be used as a brainstorming technique when you have been asked to 'talk about X', but aren't really sure *what* you want to say. The general approach is as follows:

- First, decide what your main message will be. In general this should be the answer to a question which the audience will want answered – how can we make money from this project, can we sell drug X to company Y, etc.
- Next, decide what your overall points should be. These will be the highest-level statements which will support this main message.
- Turn each of your overall points into an 'idea' statement. In other words, do not simply write 'overview of

competition', write 'product X is the leading competitor'.

- Subdivide these main points into the supporting information and extrapolations which were used to provide your statement. Note that this will generally lead to some cherry-picking of data which can support your argument – be sure that you are excluding information for a valid reason and not simply because you dislike what it is telling you.
- Arrange your 'idea' sections in a logical order, taking the sub-points with them. Then simply expand out the information.

Bottom-up structure

This is the more common situation to find yourself in, one where you have a lot of data and a number of conclusions or recommendations based on this. The storyline which ties them together, however, will often be lacking. To decide on this you will need to work your way 'up' the structural pyramid.

- You should have several recommendations, each of which is supported by a number of facts (lower levels on the structural pyramid).
- Organise your recommendations into groups based on a common characteristics – for example some may focus on cost savings, others on potential investments. These would then be two separate groups.
- Summarise the effect of implementing all of these recommendations – i.e. what would happen if all of these ideas were taken up and done. This summary will act as the start of a section which includes all following sub-points.
- Continue grouping and summarising to reach higher levels of abstraction. If you reach a point where none of

the recommendations can be further grouped, then you have reached the upper level of your report.

Clarity in writing

The most important factor in any presentation or report is clarity. You may have done a fantastic amount of research, thought up truly novel reasons for the data and developed a fool-proof set of recommendations... But it is a complete waste of time if no-one can understand what you are saying. Thus you need to be sure that your report is comprehensible for both experts in the field and busy managers.

How exactly do you do this? You need a clear and understandable introduction. You need well-written sentences and paragraphs. You need smooth transitions from one paragraph or section to another. And you need to have well-reasoned and logical conclusions. Let's look at these in a bit more detail.

The Executive Summary

Executive summaries need to pull together the main information and conclusions of the entire document. They are the first and often the only thing that busy people will read, so put some effort in at this point!

When writing, think of the reader. They have limited time and want to know your conclusions, the implications, and how you got to this point. It should not be a detailed restatement of the information in the larger document. Instead you should aim to capture to spirit of the report, not the boring details.

Executive summaries should already hint towards the questions which will be asked. Set the stage in the executive summary, ask the questions, and then restate your main points in the conclusion.

Opening the report with an introduction

An introduction is a vital part of any report, often one of the only parts (outside the executive summary) which will be read by upper management. For this reason you should put the required effort in to ensure that it is clear and persuasive.

There are three main goals which an introduction should achieve:

- It gives the reader enough background to follow the logic presented in your report
- It gets the reader interested in the following section
- It introduces the overall message or conclusion

How does it do this? There are many different approaches and the one you choose will be specific to your own personality and style. In general, however, you should think of it as a miniature story, one which tells a tale of the current situation, the looming problem, and what must be done to fix it.

The flow which this story will take is dependent on the impression you want to give to your readers. A story which begins with the current situation is calming – you aren't saying anything new, everyone reading will feel comfortable and relaxed. Beginning with the looming problem creates urgency – things are going wrong and we need to fix it now! Starting with the solution is straight-forward and clear – the reader knows exactly what you recommend right from the start.

The introduction is surprisingly hard to write, all the more so because it will have the most scrutiny from readers. It is often easiest to write the introduction last, once the rest of the report is complete, this lets you simply follow the tone of the rest of the document. You can also start on the introduction once the outline of the report is complete – though this will require more work on your part.

Concluding sections

The most important locations in any document are the start and end of each section and paragraph. A reviewer should be able to understand your proposal by reading just these parts.

Every section, chapter and report should come with a conclusion. This can seem like a waste, particularly in short chapters, but a conclusion is essential to help the reader follow your logic. A section which simply stops is unsatisfying, and will often lead to readers drawing their own conclusions about the information presented (very often the wrong one). Avoid the inevitable disagreements by including a well-prepared conclusion.

A typical conclusion will summarise the key points of the preceding section and describe how these link into the main message of the report. It should also provide some additional perspective on this information – e.g. are there any notable requirements or limitations which need to be kept in mind? Specific people or departments which should be involved?

Ensure that you have focus on important topics

Focus is vital in any written document. Your best arguments and key messages must be brought in at the start. This is not a scientific paper, where you slowly build up to an iron-clad conclusion. It is a document for reviewers, read by busy people, and you need to catch them from the beginning or not at all.

The standard rule for any persuasive document also applies to briefing books. First, say what you are going to say (and why it is worth listening to). Then say it (with data, naturally). And then say what you said (summarise the key points).

Questions should be clearly stated and have an obvious 'ask'. Try to avoid sub-questions on similar topics unless it's completely necessary – it's usually easier to read and respond when all questions are clearly separated into their own spot.

Similarly the sponsor position and supporting info should be clear and simple – not dumbed down, but shown in a way that avoid unnecessary fluff. Background information can be moved away from the questions and into introduction sections. Just don't go overboard and move the important content as well.

By contrast, information such as development history, context for indications and why you are addressing the medical need are (while important for your company) not particularly interesting for the reviewers. Move them later in the document, make sure the questions come first. Similarly information such as a manufacturing description or process overview can come later (or be skipped entirely) if not relevant for one of your questions.

Transitions between topics

Just as a conclusion helps to close off one topic, a transition is needed to move from one topic to another. It is a link between the previous topic and the next, helping the reader to remember what they have been reading while introducing the newest idea. Transitions are needed at all levels of the report, be it between chapters, between sections, or between paragraphs.

It is certainly possible to write a report without any transitions, but you will find that it makes for a very choppy and disjointed experience. This weakens your ability to get your message across, and in extreme cases your readers may simply give up and throw the report in the bin.

There are many ways to make a transition from one thought to another. For example:

- Use of words such as 'therefore', 'however' and 'thus' make for a clear transition – this sentence or paragraph is directly related to the previous one.
- Opening a paragraph with a call-back to the previous one makes for a more subtle transition. Referring to the

main point of the previous paragraph indirectly (e.g. 'this conclusion', 'previous work') and then building further links the two paragraphs without requiring an explicit 'therefore'.

- Linking two complex subjects may require an entire paragraph to be added between them as a summary and bridge between the two.

Correct use of paragraphs

A paragraph is a block of text which develops a single thought or idea. It is not a random grouping of half-related sentences (as is often seen in badly-written reports), but should be thought of as the smallest piece of logic in your report.

In general a single sentence will provide the central idea of the paragraph and all other sentences in a paragraph will relate to that thought. The main topic sentence will usually be the first one, unless you are using the first sentence to transition from the previous topic to the current one.

Avoid the temptation to mix different thoughts or ideas within the same paragraph – this will only lead to confusion and problems. Use a paragraph to develop your single idea as far as possible, then move to a new paragraph for the next idea.

The paragraph should be a self-contained piece of logic – it should include a topic, supporting data and a conclusion. It may consist of an example or analogy to increase comprehension, a comparison or contrast between information; a discussion of correlation, or a description of a group of information. The form of the paragraph will depend entirely on the subject you are discussing.

The reader will pay most attention to the start and the end of a paragraph. These are the locations where the critical information or connections should be placed, so as to ensure that the skimming reader (i.e. most of them) will remember your clever ideas. Use the start of the paragraph to link back to previous information, thus setting the stage for your new idea. And use the end to state (or even restate) your conclusion based on that paragraph's logical flow. Remember, the start and end of any paragraph are the critical parts.

Choosing words to create interest

There are many different rules for writing and many different pointers on writing style. Style is, naturally, a personal decision, and everyone will have their own writing styles. This does not mean that all styles are as good as one another – some flow like water, some will leave you scratching your head.

Rather than spend many chapters going on about rules of writing (buy another book if you are after that), this section will provide a few short tips for improving clarity. Note that many of these apply to both verbal and written communication.

- Avoid passive voice wherever possible. It is both boring to read and obscures the importance of the sentence. You can see this easily by contrasting the equivalent passive ("my mother was eaten by a tiger") and active forms ("a tiger ate my mother!").
- Boring words make for boring reading. This is particularly important when it comes to the 'typical business words'. Avoid the over-worn words such as utilise, dialogue, circle-back or optimum when you could be saying use, discuss, return to or best. As a handy rule-of-thumb, words with Latin roots are less exciting to read than those derived from English roots.
- Be careful that your modifiers don't dilute your message to irrelevance. 'Generally', 'relatively', 'mostly', largely – all of these give the impression that you are afraid to stand behind your statements.

- Don't use jargon.
- Indirect subject labels such as 'it', 'they', 'this', etc. are convenient and easy to use, but they can cause confusion if you have multiple subjects in your sentence. Indirect words should only be used if you have explicitly stated a subject in a preceding sentence and are continuing this in the next one. Changing subjects? State it again.
- Emphasise an important point by placing it at the end of a sentence or paragraph – these are most likely to be remembered by the speed-reading executive.
- Repetition of a short phrase throughout a paragraph will draw attention to the nearby words. Repetition in this manner helps to improve recall of the information so presented. Repetition, therefore, is an excellent tool to have in your writing repertoire.
- Change sentence length! Mix short ones with long ones to keep the reader interested and awake.

General writing tips

As mentioned, there is no end to the amount of things you can do to improve your overall writing ability. Here is a completely non-exhaustive list of tips and tricks.

- Balancing the amount of text you write can be quite tricky. On the one hand, you need enough to fully explain your point and arguments. On the other, too much text will cause the point to be lost. Even experienced writers have difficulty with this one, so just keep working at it, improving with each review round.
- Use plain, simple, and straight-forward language. There are no awards for clever wordplay, you aren't the next Shakespeare. Just write.
- Aim for active voice rather than passive, and don't be scared to use the word 'we'. In other words "we

demonstrate statistical significance…" is far more memorable than "statistical significance was demonstrated".

- Some words attract readers' attention more than others. Connectors such as 'thus', 'therefore', 'however' and 'but' imply an important statement, thus drawing focus to the following sentence. Other words such as 'appear', 'seems', 'in general', or 'overall' make the writer seem confident, which in turn lowers the readers trust in your statements.
- A handy rule of thumb is '5, 5, 15, 2'. More than 5 paragraphs on every page (to break up those long walls of text). Less than 5 sentences per paragraph (short and sweet!). Less than 15 words per sentence (again, concise). And a maximum of 2 commas per sentence (because you want to avoid, if at all possible, the sort of run-on sentence which drags on until the reader, by which I mean you, has forgotten how the entire thing has started, and so given up on understanding anything).
- You can branch out beyond 'standard' paragraphs of writing. Use bulleted lists, numbered items, single-line paragraphs, statements in headings, etc. etc. Anything which improves readability or information flow is a bonus when writing.
- If a table or figure is present, you don't need to restate the information in the accompanying text (unless it's truly required for the argument). But you should definitely interpret and explain what this data is saying, particularly how it relates to your company position.
- Tables and figures should have descriptive headings which flow into your arguments. Rather than saying 'Fig 3: Graphical representation of adverse events', lean into your message with 'Fig 3: Adverse events are not

increased by treatment X'. Be clear and direct the reader towards your viewpoint.

- It's also perfectly fine to put figures after the main text section referencing it. This allows you to get the entire line of thought in the readers mind before they get distracted by the image or table.
- Most briefing books will be written in English (unless targeted at a specific national health authority), which can cause problems for those who speak English as a second language. If the target health authority is in an English-speaking country (the USA, UK, Australia, etc.) then you should get a native speaker to go over the text to ensure clarity. If the authority is mostly non-natives (e.g. EMA) then they will be more accepting of 'alternative phrasing'. Save yourself some time and just send the document as written.

Creating a consistent document style

Despite all that you may have heard about not judging a book by its cover, the fact of the matter is that the presentation of a report is critically important. Readers will make a snap judgement about the professionalism of a report based on the appearance – this in turn will bias them to either accept or disagree with the findings of the author. Make sure your report looks good!

This is, however, not simply a matter of finding the 'make it pretty' button in Word and clicking it. However you will find that a few simple rules will keep your documents looking professional and reliable.

Use formatting tools for consistent style

Nothing looks worse than a document which randomly alternates between three different fonts and where headings are sometimes in a larger font, sometimes in bold, sometimes underlined, and sometimes in bright blue. Consistency is key, and the best way to achieve this is to use the built-in formatting tools which come with all modern word processors.

Taking Microsoft Word as an example, you can assign blocks of text to different formats – normal text, different heading levels, figure captions, etc. All of these can be accessed from the home ribbon with a single click. And all of these will make your work infinitely easier. Formatting can be easily changed throughout the entire document by changing the assigned format. Tables of contents or lists of figures can be automatically created based on heading levels. Including a page break after every top-level heading is a matter of two to three clicks. This saves you literally hours of scrolling through the document and fixing things up by hand.

Use headings to map out the document

Headings are the road signs of the document, they tell you what is coming up ahead and help you navigate from section to section. This is useful for both readers and writers – the writer can map out their document, the reader can follow that map.

There are different ways of structuring headings and the final choice is, naturally, up to the author. One useful approach is to require that each heading be a statement of some form – for example this section is not simply 'headings', but rather a statement that you should 'use headings to map out the document'. This tells the reader in advance what the section will recommend and helps guide them into the next section.

Having said that, many people will simply skim over a heading and move to the main text. If you have an important statement in a section heading, be sure to repeat it in the text as well. Preferably in different words.

Use hyperlinks where helpful

Hyperlinks are very useful in electronic formats, this covers both regulatory documentation as well as most reports created in the pharmaceutical industry. They are a great way to move the reader between your position and additional data in an appendix, providing a simple road map of options within the document.

However there is a very important warning attached here. Hyperlinks are inherently distracting and the reviewer may jump away in the middle of your carefully-thought-out argument. Avoid this by only putting links at the *end* of the section, not scattered throughout the paragraphs.

APPENDIX 2: REVIEW BEST PRACTICE

A briefing book needs to guide the reader towards the 'right' conclusion (i.e. whatever the company thinks is correct). It does this by blending data and interpretation from many different sources, creating an impeccable trail of logic which shows, clearly and persuasively, that your interpretation is the correct one. Unfortunately, this is usually only achieved after many, *many* rounds of review and discussion.

Document review is a major part of the workload when preparing for a scientific advice meeting. Pharmaceuticals are highly complex and highly regulated, briefing books describing their issues are no different.

But many problems arise from the review process itself. It is generally poorly managed and inefficient, with low-quality documents being pushed out the door in a rush to reach the submission date.

Very few companies have 'best practice' approaches for document review and modification. Thus everyone dreads 'that email'. The one asking for comments on the attached file, promising a dreary slog through a long document followed by a torturously slow discussion where each and every comment will be argued over in excruciating detail until finally…

You get the idea.

This chapter will go into the review process – what works, what doesn't, and how you can improve your actions within the company.

What exactly is a review?

'Review' is a fairly broad term and can be taken in many different ways. In this book we focus on the idea of 'reviewing' as analysing the document at a *strategic* level. This means you are reading it to see if the text fulfils very high-level requirements:

- Does it get the required message across?

- Does it achieve its purpose?

- Does it meet the needs of the target audience?

This is different to other processes which are also commonly called 'review'. For example, you may have to assess the *final* document after it is completed (think of 'peer review' for a scientific paper) or verify that the document meets the required rules (are company templates used correctly, is the data correct, etc.), or check that the spelling and grammar is correct.

All of these are important, but *nowhere* near as important as the strategic-level review. The FDA will overlook an ugly document format, they won't overlook a nonsensical argument.

Review also needs to be a useful and co-operative process. The reviewer needs to point out problems as well as possible solutions, and they need to do this alongside the document writing process. In other words, it needs to happen as you go, not in one big lump at the end.

The usual review process doesn't work well

The general review experience is not a good one. We've all seen it happen. Inefficient and repetitive rounds of discussion and comments, the entire process taking up valuable time and often leading to a sub-par final document. This occurs for a number of reasons, but the main problem is that reviews come *too late* in the writing process, and focus on the *wrong things*.

Very few authors manage to set up a clear review process with defined goals for each stage of the document writing process.

Instead they will send an email out to a long list of people, draft document included as an attachment, and then see how everything goes. The reviewers will be selected based on political reasons, for historical reasons, or because they have always been on the mailing list and no-one can be bothered removing them.

If you are lucky, those involved will be experts in a particular area under discussion (clinical, analytics, etc.), reviewing as a subject representative. Later reviews generally involve higher management functions of those areas. Upper management almost always gets involved in the last review rounds. Their job is intended to be a focus on the strategy involved, but the majority of their comments will involve language use, document structure, or useless complaints (e.g. 'I don't really like this part').

The reviewers themselves will open the document, sometimes remembering to save a version with their initials on the end. They will then work their way through the text, start to finish, making comments on anything which catches their eye. They generally won't discuss their comments with anyone, but they will send the commented version back to the author.

The author will then collate all of the commented versions together, somehow finding a way to merge them into one document. They will hold a few round-table meetings where the document will be discussed, the entire group moving down the document from start to finish, arguing as they go. The conversation will inevitably be side-tracked by a discussion about the scientific merits of one minor point and the group will only get back on track a few minutes before the meeting ends – at which point another meeting will be scheduled.

Does this sound familiar?

Here is a list of common problems during review:

- **Lack of standards**: There is no fixed approach for developing and reviewing a briefing book. This means that the participants are not clear as to what they are doing and by when it should be done. Each reviewer will also have their own idea of what the 'required quality' is, leading to wildly varying results.
- **Timing problems**: The writing stages and review rounds often don't line up, which means that the key people get involved too late or the authors begin writing before the team has agreed on the correct approach. This leads to wasted work and multiple review rounds.
- **Reviews come too late**: Following the previous point, reviews inevitably occur late in the writing process once the majority of the work has been completed. Changing the strategy is far more difficult at this stage and even minor comments can lead to significant rewriting.
- **Irrelevant comments**: Every review round will have many people reviewing and commenting on different sections. Comments will often focus on minor things such as grammar or writing style rather than strategic issues. This creates a sea of distractions which the author has to wade through in order to find the important comments.
- **Conflicting comments**: One set of reviewers wants one thing, another set wants another, and all of them change their minds each time the document comes back for another review round. This is particularly bad when dealing with different management levels.
- **Time pressure**: There is a huge amount of pressure to get the briefing book done and off to the health authority as quickly as possible. This leads to rushed work and short-cut solutions to complex problems, while those

leading the process rarely have a chance to tell the others what they should be doing.

- **Focus on traditional approaches**: Reviewers place a lot of importance on how things were done in previous briefing books or health authority interactions. This prevents the authors from trying new methods which are more suited to the current problem.

Performing a good review

Enough of the current situation, what *should* be happening when people go to review an important document?

A good ground rule is that all reviewers should be focusing on the *strategic* aspects of the document. In particular:

- Are we presenting the right message?

- Do we have the best arguments?

- Are those arguments supported by the data?

In addition there are some basic guidelines to follow when organising the review process:

- **Organise the process**: Make sure you plan out the review and writing process so that everything flows smoothly and according to timelines.
- **Get the right people**: Get the critical reviewers involved early in the writing process, and make sure they have clear roles and responsibilities.
- **Review early and often**: Early reviews need to focus on the scope, message and arguments which will be fleshed out as writing continues. Even if the full details aren't there, you can still get an idea of whether the approach will work or not. In general you should be having reviews of the outline, early and late draft stages.

- **Write better comments**: Make sure the comments you write are useful for the writer and bring the document to a better state
- **Have consistent standards**: Make sure everyone involved has a standard definition of 'quality' and 'document requirements', so that the reviewers will pick up on the same things.
- **Discuss comments intelligently**: Prioritise reviewer comments and spend time discussing those which actually matter, rather than treating them all as having equal importance.

These guidelines are discussed in further detail in the following sections.

Organising the process

As mentioned before, being able to plan out the writing and review process is a vital part of getting everything done quickly. But once a plan is made, you need to make sure that everything goes according to it – something which is not easy even in the best of times. To avoid chaos, you need to think of the review process as something which needs to be *managed*.

In most organisations the planning and organisational responsibilities will fall on the lead medical writer or regulatory affairs manager. Sometimes a project manager will be involved, but it is generally best that whoever really *does* the work is the one to organise the reviews as well. They know the issues and the solutions best, adding extra people into the mix just muddles everything up.

The responsible person will need to ensure that the right people are chosen, that drafts are sent out, that comments are compiled, and that reminders are sent to that one manager who never checks their email. A nice summary of their overall responsibility during organisation would be as follows:

- **Identify the reason for the review**: First you need to identify the purpose of the review – why do you want people to look at your draft? ('Because my manager told me to' is not a valid reason here). Which areas of the document are weak and need discussion? Which are still early and awaiting further data, and so don't need to be discussed or criticised? Are they reviewing at an early, strategic level, or providing their opinion on your late-stage draft?
- **Plan out the overall timelines**: Make sure the overall submission plan includes sufficient time for reviewing, meeting to discuss comments, and implementing the decisions. It is often helpful to get this included in the main project plan as well so that everyone is aware of the timelines.
- **Choose the right people:** Don't just choose any old reviewer, make sure that they are necessary for the review round underway.
- **Tell the right people what to do**: Get the document draft sent to those people on time. Ensure that instructions for reviewing are included as well so that they are more likely to follow best-practices. It helps to include a checklist for the review to focus their attention on certain factors, as well as formalising any previous experiences.
- **Provide specific instructions**: You should avoid simply asking for a 'review', as this is vague enough to leave your reviewers slightly confused and ready to fall back on their instincts. Instead, any request for a review should ask people to focus on a specific issue, for example the setting of specifications or the interpretation

of a clinical readout. Your goal, essentially, is to set the stage for your reviewers by framing the requests and targeting those requests to the skills of the reviewers.

- **Provide context**: Reviewers may need more information than you have available in the draft. This is very often the case when background knowledge or recent information needs to be taken into account. A simple example is following implementation of a new guideline or health authority requirement – this will be clear to the Health Authority that reads the briefing book, but perhaps not so clear to your manager. Any additional information should be provided to the reviewers at the same time as the document for review. It will help them to understand the context of what they are reading, which in turn means that their comments will be more helpful
- **No really, provide specific instructions**: Make sure everyone involved is clear on what they should do and what is considered in or out of scope. It is also often necessary to explain the overall purpose of the briefing book to reviewers who are new to the topic (this is a typical problem when dealing with upper management reviews).

Getting the right people

There is no need to invite all reviewers for all review rounds – in fact this is the least effective way of going about things.

Instead you should target your experts to the document stage. Scientific experts should come in early when the data is being populated and initial conclusions are being made. Upper management and regulatory affairs will get involved to ensure that the conclusions and supporting data meet the needs of the company and the upcoming submission. Finally QA and QC will do the final polishing and data integrity checks. You may also wish to have external reviewers come in and do an independent 'reality check' of the briefing book to ensure that it actually meets requirements.

Remember that it is *always* easier to invite new reviewers in than to get old reviewers out. In particular, people will fight tooth-and-nail to remain on the review list when they think that money or prestige is on the line.

The reviewers themselves should bring their knowledge as a subject matter expert to the review process. But they should not review *as* an expert, rather they should review as though they were a health authority reader with a similar level of expertise in the field. In other words, you should think like a clinical expert from the FDA, not like the clinical expert from a pharma company.

Reviewing early and often

Once the document plan is in place, you can begin to organise the different reviews which will occur at the different **stages of development**.

What does this mean? Review is not a one-size-fits-all process, but rather one which focuses on different factors at different times. By explicitly setting different goals for each stage of the document preparation process, we can force reviewers to concentrate on the relevant factors for that stage. This is detailed in the following sections, but the basic approach is to focus on overview topics at the beginning and move to more detail as the document nears completion.

This approach can be challenging – authors in particular need to stop thinking of drafts as 'near final' and accept the idea of providing early-stage drafts to a potentially critical audience. Few like the idea of their ugly, early work being exposed to reviewer comments. Even worse, some of the *reviewers* may not understand the staged review process, instead thinking that the early-stage document is just a badly-written work from someone that we really should fire for incompetence...

This just underlines how important it is for all involved to follow and *understand* the staged review process. In general you will want to have SOPs or equivalent instructions in place to set expectations. For example, you may have an SOP covering the requirements for an early-stage draft, as well as a list of which experts should be reviewing and what they should be looking for.

You should also have a list of what experts are *not* allowed to comment on – people should not be arguing about quality-control or data integrity factors at the first draft, these should be checked at the end. It helps to include a strict naming system so that everyone is clearly shown from the document title that this is, for example, a preliminary draft or an outline.

The early review stages in particular are helpful for determining the strategic direction which the document will take. They should usually be reviewed by upper management, especially those who will have to endorse the findings of the document at a later stage. Let the different parties have their political arguments now, while everything is easier to modify, rather than later when the document is near-final.

Review goals for review stages

So you are on-board with the idea of reviewing at different stages? Great! Now to figure out what everyone should be looking at.

We divide the document writing process into four main stages, the Outline, the Early Draft, the Late Draft, and the Final Draft.

- The **Outline** is a very rough look at how the data and arguments will eventually be placed in the document. You may hear it referred to as a 'shell document' or a 'seed'. This is where the eventual chain of logic is laid down for later use.
- The **Early Draft** has text and data, but it is still a clear work in progress. At this point you discuss the overall organisation and strategy as it is being put together.
- The **Late Draft** looks like a real briefing book, albeit with a few rough edges. At this point your reviewer comments should focus less on strategic issues and more on persuasiveness and completion.
- The **Final Draft** is basically finished, it just requires polishing to ensure that everything is as impressive as you can make it. There are often several rounds of minor changes at this stage, and you can often expect to receive a briefing book named "Document-Final_edited_FINAL-v2_Final.docx" before it is over.

You may have different names for these in your organisation, this isn't really important. What is important is that you direct the review process according to the stage of the document – a reviewer should be looking at different factors in the earliest outline versus the final document.

This section will cover a few of the main points you should be looking for at each stage.

Outline

At this point the document is just a sketched outline of what it will eventually become, you may even find it is nothing but headings and a few boxes to represent figures. Despite this, decisions made at this point will have long-reaching consequences, so you will need to take care with the reviewing and designing process.

The outline should show the layout of the final briefing book, including the different section headings and tables and figure (though the data itself is not necessary at this point). It allows the team to define the overall framework of the argument and the critical information which will be presented.

In particular, it helps to look for the following:

- **The core message of the document**: Have you defined the core messages which the document will deliver? Does everyone agree with these messages, so that they can all work together to bring them forward? Does the proposed framework make sense for the argument you are bringing, given the data we intend to include?
- **The interpretation of the data**: Given the data you have, what is the most important set? Should it be presented as a graph or separate table to provide more prominence? Is the data which will be included appropriate and able to support your strategy? Does everyone agree on the message presented by the data? (**Note**: This is the point where you should have the big arguments about interpretation, not at the later stages)
- **The underlying studies which were performed**: Does the study design makes sense? Is it based on sensible rationales? Do you have sufficient data to write up the study? And are your interpretations of the data correct? Has the study followed the correct research and development processes? Relevant SOPs, GxP, etc.
- **What could go wrong**: Have you identified the potential issues with the study and data? If so, have you aligned on a response to these? How will your proposal and the potential issues impact on the eventual regulatory submission? Have you discussed the eventual effect on downstream processes such as the label, marketing, or manufacture?

- **The overall look of the document**: What document style will you be using? How will the headings, fonts, tables, and the like look?

Early draft

The early draft will begin to look like an actual document. Text sections will be blocked in, data will be in tables, you will start to get a feel for how it will look at the end. At this point you need to check whether the planned document will meet your planned requirements.

This review stage focuses on the arguments used in the document. Are they strong arguments, and do they work with the data at hand? It is also a good point to start checking accuracy and completeness of source information, as well as consistency across the briefing book.

More specifically, you should check:

- **The chain of logic**: Look at the key scientific arguments being drawn from the data. Do they make sense? Are they clear? Is there a clear line from the data through to the claims you are making? Do all of your claims fit together, or are some contradicting each other? This is closely related to…
- **Key messages**: Are the key messages which you defined before now in place? Have you placed them in the right sections, in a way which is clear and easy to understand?
- **The gaps in your argument**: Is your data strong enough to support your claims? Do you need to perform additional studies? Are there gaps in the arguments or mismatches between data and interpretation which the health authority could jump on during the meeting?
- **Document layout**: Have you got the organisation of the document right? Is the right content in the right place,

easy for the reader to find and highlighting the important findings? Is it a useful layout or merely serviceable? Are you following the 'standard logical layout' of a scientific document, or a more sensible approach for busy readers?

- **The problems**: Are the issues you identified earlier clearly and comprehensively addressed? Have you included enough data or argumentation to close them out? Enough for readers who *don't* have your level of background knowledge on the project?

- **Regulatory requirements**: Does the information present meet the relevant regulatory requirements for your briefing book? Have you addressed any feedback which you may have received beforehand? Is there anything which the health authority will expect to see that is missing?

- **The tables and figures**: Now that your graphs and tables are filled in, do they deliver a clear, unambiguous message? And is it the message you want to deliver? Do they make sense and clearly answer a point raised in the main document? Can you understand them, finding the important data right away? And if there are outliers or strange results, have you explained them?

- **A strong conclusion**: Do you have a strong discussion and conclusion for each point? Are the objectives covered, with well-argued logic in place? Do you have a real *conclusion*, or are you just summarising the results?

The logical layout of a scientific document

A briefing book, like most scientific documents, has a fairly consistent layout of logic. The main difference is that a briefing book heavily focuses on the data and interpretation for specific issues, rather than providing information on everything which was done (this is the job for the appendix).

More specifically, you will generally see an 'aim' to the study, this tells you why the research was done in the first place. This feeds into the overall conclusion of the document – we wanted to do it, so why are our findings important? And what do we do now?

The goals or objectives state specifically what was done to meet this aim – specific targets which can clearly be quantified and 'ticked off' when successful. The initial targets are then closed out in the later discussion section, which justifies the importance of your results and whether they met your objectives.

The parameters which you measure are chosen to meet the stated objectives. They are (hopefully) defined before you do the experiment and the results then shown in the document. A mismatch between proposed parameters and reported results looks like you are hiding something, and this is guaranteed to draw questions from the health authority.

When writing a briefing book you will need to summarise this information and focus on the most relevant factors. As there are usually multiple questions in each briefing book, you'll also see some repetition and 'mini-arguments' within the background to each question. But remember that reviewers are also scientists, and so will expect to see a similar chain of logic in a briefing book as they would expect in a more detailed document.

Late draft

Things are getting more serious now, the document should be almost fully fleshed-out. This review round will focus on the persuasiveness of the document, ensuring that it will support your required message.

One of the biggest tasks at this point is to ensure that all comments made before are addressed in this version. In the high-pressure and rushed writing process it is entirely possible that some critical points were missed – checking this is fairly simple and will save you a lot of trouble later on.

In particular, watch for:

- **Real discussion of the topic**: Is the document more than a simple list of results? Does it go beyond this to actually make conclusions about the findings and their significance?
- **Major messages are in place**: Does the structure of the document provide emphasis and persuasive commentary on all major points? Does it argue against any problematic issues? Does each section have a strong beginning and conclusion to lead the reader in the right direction?
- **The data supports your conclusion, and nothing more**: Double check that the data and analysis are accurate and scientifically justifiable. Are you absolutely sure that the data supports the conclusions? Are there other interpretations which the reader could make? Think like a regulatory assessor – what holes could they possibly poke in your arguments? How can you prevent this?
- **Interlinked text and diagrams**: Are your tables and graphs are self-contained, in that they can be understood without referring to the surrounding text? Despite this, does the surrounding text reference and expand upon them? Are they well-designed and -labelled?
- **Persuasive writing style**: Check the writing style. Is it clear, correct, and persuasive? Do your sentences make sense? Are you precise and consistent when referring to information?

Final draft

The document is almost done, which means that reviews at this stage will focus on the polishing of the final product. Much of this work will seem boring or picky, but it every bit of additional polish will help persuade the reader that your solution to the issues at hand is the correct one.

What does this mean in practice? For example:

- **Strong introductions**: Does the start of every section clearly outline the content of the section in a way which supports your main message? Can you understand the main arguments simply by reading these introductions?
- **Strong transitions**: Does the starting sentence of each paragraph spell out the topic of that paragraph? Is there a smooth transition from one paragraph to the next? Do you feel that you are flowing along the chain or arguments, or are there points where you need to stop and think about things?
- **No errors**: Have you done a quality control check of what is in the briefing book? Are you absolutely sure that it matches up with what is in the source documents? Do a final check of all figures and tables. Are they all consistent, legible, and in the right places? Does all of the text follow the set style convention? No typo errors or missing references? Have you double-checked if the title page and signature page are correct? You would be amazed at how often this is missed.
- **Persuasive arguments**: Are you sure that the problematic issues are covered by data-based arguments? Do you find yourself nodding along to the claims made by the author? Are you completely sure that the regulatory requirements are fulfilled? If someone paid you a million dollars to poke holes in the arguments, could you?

So you're a reviewer

Much of the previous sections have focused on the role of the organiser – getting document drafts out, collecting comments, organising the discussion in productive ways. But sometimes you will also find yourself as the reviewer, which comes with its own set of skills and difficulties.

This section covers the important 'ground rules' of being a reviewer, then follows up with a number of techniques which help you to spot issues quickly.

Ground rules for reviewers

There are a few basic ground rules which you should keep in mind when acting as a reviewer. The following list covers pretty much all of them:

- **Follow the instructions**: If the author gives you instructions on what to review and what to look at, follow them! Don't go off on a tangent. You should already understand what the document is for and how your speciality fits into the writing/review process – so keep your review focused on what they are asking for. And if you haven't received any instructions, then ask for some!

- **Read it multiple times**: The most successful review approach requires around three runs through the document. First you skim the document to get an impression of the logic and arguments. Then you read it in more detail to see how it presents the message, following topics that interest you through the document via hyperlinks or search. Then you read it once more from start to back, adding all the comments which you have thought of in the previous run-throughs. Why do this? It pulls you out of a linear approach and forces your attention onto overarching topics – providing more strategic feedback.

- **Provide useful feedback**: Comments like 'redo this part', 'I don't like this' or (worst of all) '?' are useless. Provide comments which explicitly describe what needs to be changed and *why* so that the author can figure out what to do.
- **Read like a health authority reviewer**: Look for arguments or data which don't seem particularly convincing. Tell the author where you see gaps. If you were confused, say so. If you think there is an alternate interpretation, say so. Remember that if you are confused by something, then the health authority reader will be too.
- **Challenge the document**: Don't let yourself get dragged along by the story of the document, actively think about why they could be wrong. Read along as though you are a sceptical and slightly nasty expert in the field. Look for the holes and challenge what is written. Are they completely wrong? What data would you expect to see which is missing?
- **No editing**: Editing on the word or sentence level is a waste of your time. You are here to give strategic feedback and improve the quality of the document at a high level. If you have general comments on the language quality then yes, go ahead. But don't add commas or change wording – add a comment if you have to, otherwise leave it alone. Editorial comments should be made once and then disappear, never to be seen in a further meeting.
- **Use a time limit**: Set aside a chunk of uninterrupted time (usually 2-4 hours, preferably split over several days) where you can sit down and really focus on the review. Don't get distracted. Sometimes it helps to split

the different review passes over multiple days so that you don't get bored of the document.

Reviewing techniques

There are many different techniques which have been developed to improve reviews, and we present an overview of the most useful ones in the following sections. Whichever technique you follow (or whichever you tell your reviewers to use), you must ensure that all reviewers know why they are reviewing the document, what they are expected to provide, and what steps they will need to follow.

In other words, don't assume that everyone knows what they are doing.

Checking organisation of information

The manner in which your information is presented and organised is a vital part of reader understanding. A confused reader is one who doesn't agree with you, after all, and the main aim of a briefing book is to get health authority buy-in on your proposal.

First, verify that all information in any particular section is relevant to that section. How do you know if it is relevant? It should fit under the topic as defined by the heading. If it doesn't fit, you either have information which should go somewhere else, or your heading isn't summarising correctly. In either case you will need to fix something.

Next, ensure that any information presented in a higher-level section summarises the information in the lower levels. Think of it as a pyramid, where details provided in the tables are progressively condensed into summaries as you move up towards the main conclusion. Alternatively, the document becomes steadily more detailed as you work down through the layers of headings. Bringing in details at a higher level will confuse the reader, failing to summarise will leave gaps for them to pick on.

Review the key positions

The 'key positions' are those which a normal reader will focus on when skimming the text – they are also sometimes called 'positions of prominence'. In a typical scientific document these positions are the first paragraph under a heading, the paragraphs immediately before and after a figure/table, and the heading of the figure/table itself.

You should be able to pick up the briefing book and understand the overall argument by reading only these key sections. More importantly, a skimming regulatory reader is almost certain to cover these positions, thus any mistakes here will certainly be picked up. When reviewing, check that these sections are complete, well-written, and consistent with the overall message of the document.

Verify the starting sentence

A well-written document will split the overall argument into discrete chunks of logic and information, these will then be wrapped into individual paragraphs. And a well-written paragraph will always start with a key sentence, one which introduces the topic and argument of the following sentences. Thus by reading *only* the first sentence of every paragraph, you should be able to understand the argument being made and the manner in which the points flow together.

This is a difficult review technique to do as we will automatically begin reading the suequent lines. Nonetheless it is a powerful technique, as it provides a summary of the entire logic and sequence of discussion as though you to a new reader. One approach which can help is to copy the sentence out to a separate document file, though obviously this is quite time consuming.

Keyword assessment

Regulatory readers are more and more likely to use search functions to find the keywords they are looking for. Thus you need to do the same.

Use search to find keywords which you would consider relevant to this document – for example a clinical study briefing book would likely include a discussion on study power. Then check each instance of the keyword to see what context it occurs in. If you cannot find this keyword via search, or if you cannot find the expected information, then you can expect the regulatory reader to have the same difficulty.

Writing better comments

Many reviewer comments are, well, kind of useless. Or to be more polite, they are simply difficult to understand. You have inevitably seen several examples – think of the inevitable "?" or "rewrite this" or, much worse, "make this pop more". These comments don't help the author and often stray into the copyediting field, which is usually the least-useful thing that a reviewer can be doing.

Instead, reviewers should focus on a number of broad, overview-related questions. Namely:

- Does the document meet the overall purpose?
- Would it support our needs in a regulatory or quality-related matter?
- Does it fully address the most important issues?
- Is the most important information in an obvious and easy-to-find location?
- Can the reader easily navigate through the document?
- Can the data tables and figures be understood as they are, without referring to the text?
- Do they make clear and well-supported points?

Ideally you will cover each of these questions on a different pass through the document. In other words, do not simply go page-by-page and comment on whatever you notice, but really think about certain topics as you work your way through the document.

The most useful reviewer comment will note that a problem exists, describe why it is a problem, and then offer a solution or suggestion to fix it. This solution may not be ideal, but it should at least get people discussing in the right direction. The comment should be in full sentences, rather than cryptic scribble. And it should offer suggestions but leave the final wording to the author – this prevents reviewers from arguing over minor issues such as commas and synonyms.

One nice way to force this is to create a review form which is provided to all reviewers. As part of this, include a section which specifically requests comments on the document as a whole. This sidesteps the usual problem of having all reviewer comments in the margin of the document, which automatically makes them focus on discrete sections rather than the overall piece. Despite this making global comments is a rare skill and one which reviewers will need to practice and train in before reaching any sort of mastery.

It also helps to split the editing and review process. This can be done by designating one reviewer as the copy editor or even bringing one in from an external source – their job is to fix all of the grammar and wording issues which may otherwise get in the way and thus prevent many arguments over minor errors. Try to ensure that only one person does this, as this keeps the grammatical and writing style similar throughout the document. You may also want to perform the copy-editing pass at an early stage in the document preparation to remove the most obvious errors, followed by a final correction prior to finalisation.

The entire process needs to be thought of as a collaborative one. You aren't trying to catch out the author's mistakes, you aren't aiming to make comments that seem clever but aren't particularly helpful. All comments should be designed to bring the document to a better state, and should be brought in early enough that improvements can be made with a minimum of difficulties. The overall success of the briefing book is everyone's responsibility, after all.

Using software tools

There are many different tools which are available to speed up the reviewing process. As there are any options it is possible to choose one which suits your preferred style of reviewing or chosen company method.

The main advantage of dedicated reviewing software is that it avoids the typical 'attachment dance' – a process by which the author will email out a copy of the document to multiple people and then collect the individual replies in one master document. This forces every reviewer to work in isolation and provides an exceptional amount of decision-making authority to the author, as they alone are able to decide if comments should be integrated into the main version. It also tends to accumulate many, many versions of the same document, often with names such as briefingbook_final, briefingbook_final_v2, briefingbook_final_v2_reviewed, etc.

It is important that reviewers be able to see each other's comments. This lets them work through issues and ideas amongst themselves, presenting alternative ideas and solutions without the need for a dedicated meeting. It also provides recognition for those reviewers who make useful, workable comments and encourages the rest to improve their game.

Commenting tools exist in most modern software systems. Many of the most common systems (e.g. Word) treat comments as being placed 'on top of' the document – in other words the original text prior to comments or correction can always be seen, allowing you to roll-back to the earlier version as needed.

Interesting the tool being used does tend to make a difference to the people using it. Paper printouts are often preferred by people doing quality-control checks, auditing, comparison between documents or between related-yet-separate pages in the same document. This is usually because you can spread out different paper stacks across the entire desk, while there is a limit to how many monitors the average worker has available.

Having consistent standards

The entire point of having a review session is to find problems and suggest improvements. You are not looking to see if it is 'good enough' to send, you are actively trying to improve the document from its current state.

The most important review is that which occurs at the strategic level, as described in this chapter. Writing style and grammar checks are important (the eventual reader won't believe you if they can't understand you) but they are generally *much less* important than a strong argument and persuasive data-set.

This is why review should happen at multiple stages during the writing process, with a heavy focus on these strategic questions – does it make the right arguments based on the right data?

But a review is useless if everyone is pulling in different directions and providing mutually exclusive comments. So all the reviewers need to be on the same page when it comes to expectations – what do we want to have as a working style, what is considered 'good' writing, what makes a strong argument, etc.

Working style is dependent on the organisation you have and the culture that exists there. But generally everyone should remember that it is a *collaboration*, not a competition. The reviewers are not trying to catch out mistakes because the writer is a moron, they are trying to make the document better by working with the writer. Any other process leads to anger and stress and, eventually, dries up any open communication.

At the same time there should be consistent standards for the reviewing process. By creating a consistent procedure for the review you not only speed up the entire process, but ensure that experience from previous attempts can be brought into the next one.

A helpful addition here is the use of checklists to guide the review process. Checklists are a handy way to keep everyone on track – you can even use specific checklists for different departments to avoid cross-reviewing. By reviewing the list each time a briefing book is submitted you will slowly accumulate 'organisational knowledge' about what should and shouldn't be done. This then makes the next time at least slightly easier.

Discussing comments intelligently

By this point the document should have been sent out to all the relevant reviewers and they have (hopefully) provided a series of intelligent and thoughtful comments in return. Comments will range from breathtakingly important through to incredibly minor, and it is very important that you are able to distinguish between the two extremes.

Once you have all of the comments back, it is time to compile all of them together into one giant list. You will find that many of the points overlap or repeat themselves, allowing you to identify common problems and condense the large list down to something more manageable. This compiled list should then be organised based on the importance of the comment being made.

- **Highly important** comments point out things that will prevent the document from succeeding. This can be missing content, unresolved issues, completely wrong conclusions, whatever it may be. Because these are such a problem for overall success, you need to find a solution with the entire team.
- **Medium importance** comments discuss areas which are confusing to the reader or which are overly ambiguous. They don't prevent the reader from figuring it out eventually, but it takes them out of the flow of the document and makes the overall argument less believable. These are best solved through direct communication between writer and reviewer, and rarely need the full involvement of the team.
- **Low importance** comments are those regarding style, grammar, minor data errors, etc. These can be taken up by the writer and simply fixed directly. You do not need to bring up this kind of comment in any further meeting with the team or reviewers, it simply isn't worth the effort. Having said that, you really shouldn't be seeing comments with low importance when reviewing early drafts – if you do, then it's time to remind the reviewers of their instructions.

Once all comments are organised and prioritised, it's time to discuss the important ones with the rest of your team.

The round-table review meeting

Basically every company will follow up the reviews with a round-table meeting, a face-to-face meeting where reviewers and the author will go through the document under discussion. These are, to put it bluntly, rarely useful.

The typical process involves everyone assembling and going through the document page by page, from start to finish. This is the point where you will discover that many have not really read the document and almost no-one has looked at the comments of the other reviewers. There will be a half-hour long discussion on a single major topic, which won't be resolved in any meaningful way. The meeting will then descend into petty arguments about grammar and punctuation.

Alternatively, and more intelligently, you can take a structured approach. By planning ahead and managing the meeting process it is possible to bring actual results out of the usual chaotic free-for-all.

The first and most important step is to *get the prioritised and compiled list of comments* in place at an early stage. This is critical to keeping the meeting focused and can be the most difficult part to achieve. Constantly nag your reviewers to get the feedback, particularly those who you know are always late. Lie about the deadline if you need to. People who only bring up a problem at the meeting itself are hugely disruptive to the rest of the team, so make sure they don't do it.

Once you have the list, you should *provide the comments to all meeting participants*. Round table meetings may involve just the team (the standard 'oh god how do we fix this gap?' meeting) or may include both team members and reviewers ('do you believe we've fixed this gap?'). Regardless of the composition, all participants should have the latest version of the document to work on. It helps to *include guiding instructions* here – create a list with the major issues so that the meeting focus is clear, mark minor comments as something to be done without team involvement, and demand that every participant reads the full document and all major comments before the meeting.

The meeting itself *should have a defined agenda* covering the high priority comments. It will describe the points for discussion (most important ones first, obviously) and propose a time period for discussion and solutions. This can be flexible – you should be find to go overtime if the discussion is running well and the team is coming to an answer. But you shouldn't let one reviewer comment dominate the entire meeting – if it won't be solved in a reasonable time, then either delegate to a sub-group or call a follow-up meeting just for that comment. This keeps everyone on track.

At the same time you should *follow standard meeting best practice.* Make sure you have a facilitator to run through the agenda and keep the discussion on track (this is usually the job of the main author). You should also have a separate note-taker to write everything down – it is very difficult for the facilitator to do both and inevitably leads to a lot of forgotten ideas. The facilitator will need to keep the team focused on the topic at hand, away from anything grammar or style related, and ensure that no-one tries scrolling through the document from start to finish. You want to be focused by priority, not sequential-order.

Once the meeting is over you will (hopefully) have solutions to the tricky problems the reviewers brought up. Now is the time to summarise the discussion in a set of meeting minutes and make sure all are clear on the agreed approach.

Once this is finally done it is now just a 'simple' matter of implementing the agreed-upon changes into the document.

Easy, right?

APPENDIX 3: PROBLEM SOLVING STRATEGIES

One of the most important skills which any one in the world of science can have is that of problem solving. Everyone on the project is a problem solver in one form or another. But not all of them are good at it.

So what makes a *good* problem solver? You need to be able to provide innovative and useful insights into a wide range of situations, combining those into solutions which are fact-based, clever, and pragmatic. This may seem well beyond you but, like any skill, problem solving is something which you can learn, practice, and master.

Problem solving tends to rest upon a few main requirements:

- A well-thought-out definition of the problem.

- A process for analysing and solving the problem.

- Flair and creativity to go beyond the standard solutions.

- A willingness to accept help from others.

Because this is such a complex area, there are many, many other books which go into problem solving in detail. We'll provide a basic overview here, but feel free to look at other books in the Further Reading appendix.

What makes a good problem solver?

It is often difficult to say just who will be a good problem solver – many times the best idea will come from the people you least expect to hear it from. However there are several innate or learned talents which good problem solvers will possess:

- Able to see a problem from many different angles

- Able to see connections between different elements, both inside and outside of the defined problem
- Able to find the 'main point' of a very complex subject
- Able to look at the broader implications of an otherwise targeted decision

If a you find someone with these talents in your project team, try very hard to hold onto them!

Steps in problem solving

There are many ways in which we can look at problems and attempt to solve them. Generally they require a number of steps, in which 'analysing the problem' is only *one* of these steps. In other words – what you think of as the most important part of problem solving is actually just a small part of the whole process.

Effective problem solving requires that you move from the initial step (wondering what the problem really is) through to the last part of the project (recommending a solution).

Briefly, these steps are as follows:

- **Define the problem**: Possibly one of the most important steps, as you cannot answer a question without knowing what the question is. Defining the problem is the act of deciding just what exactly you need to solve.
- **Assess the problem**: Look at the problem you have defined, and try to spot other problems within it. Can the problem be broken down into sub-problems? Do early hypotheses spring out at you?
- **Prioritise**: Determine which issues relating to the problem should be considered most important. Don't forget to determine *why* this would be the case.
- **Plan the analysis**: Plan out the areas you will focus your investigation on and how your analyses will be

performed. What is the most efficient way to spend your time?

- **Conduct analysis**: Analyse the data you have available. Try to focus your work on the priority areas you have identified – what hypothesis are you trying to prove or disprove? How does the evidence relate to this?
- **Combine your findings**: Bring the information which you have gathered together and try to find the common thread. This is where you ask yourself what implication the potential solutions would have for both the problem and the company as a whole.
- **Make a recommendation**: Your work is useless without a final recommendation. What should the next actions be? What will this achieve?

Some tips for defining the problem

The first step in solving any problem is to successfully define the problem. This should follow the usual SMART rules – it should be specific, measurable, action-oriented, relevant, and time-bound.

Many use a 'problem statement worksheet', which is realistically just a nice-looking checklist with the important factors you need to include when defining a problem.

These include:

- **The basic question to be answered**: The defined problem, according to SMART goals. The definition should be broad enough that the important external factors are captured, but not so broad that you become mired down in irrelevancies.
- **Perspective/context**: This section sets out the situation that you are in and the complication which is affecting them

- **Criteria for success**: You cannot attempt to solve a problem without knowing what your victory conditions are. This section should define what precisely the success criteria will be, something which will need to be jointly agreed upon within the project team. This will often take a while to agree on. The success condition should be both qualitative and quantitative, and should allow you to clearly point to the moment when you have solved the problem.
- **Scope of solution space**: Basically what elements and possible variables will be included in the project. Often more important (and usually easier to define) is the list of what will *not* be included.
- **Constraints**: What are the limits or constraints to your potential ideas? What options can't you consider, even if they may seem good? These are usually driven by budget, regulatory, or political reasons.
- **Stakeholders**: Note down the important people involved in the problem. This includes the people who will make preliminary or final decisions on whether to accept your recommendations. You should also note who has the ability to support or block the project – this is usually management but may also include front-line workers or people whose power or income are on the line.
- **Key sources of insight**: This is a listing of the places you will go in order to get information on your problem or potential solutions. This includes databases, health authority guidance, web searches, known experts, previous projects, advice from mentors, etc. etc. By formally writing this down you ensure that no source will be forgotten in the mad rush to get everything finished.

Tips for creating alternate perspectives

It is important to view any problem from a number of different perspectives. Coming in with a single, fixed viewpoint will lead to you automatically discarding any number of potential hypotheses based on your own pre-set beliefs.

This is, obviously enough, not a good thing. The more options you examine, the more hypotheses you can generate, the more valuable your eventual recommendation will be.

How do you improve the number of viewpoints which you can bring to bear on the problem? There are several different approaches you might want to try:

- **Try on other people's shoes**: This sounds rather cliche, but one of the most basic ways to find new viewpoints is to think of how other people would react to the situation. Take any of the stakeholders and think what *they* would consider to be important factors – in general this would be things which directly impact on their own self-interest. If you take *those* priorities and use them as a basis to look at the problem, what stands out as the most important aspect? Why? Now repeat this for the other stakeholders involved – department heads, regulatory authorities, project team, even the competition.
- **Look at a broader context**: The problem you are looking at is usually defined as affecting a single product, group, or company. But can you expand the context to look at more options? Does something *similar* affect other products, companies, even the entire field? If yes, what are the similarities?
- **Relax your constraints**: Any problem definition will need to include a description of what is not in scope. But what happens when you take some of those constraints away? Are there factors which you have been told not to

assess but which are nonetheless relevant? You need to be careful here – although the client may have missed some important context, it is also possible that they are well-aware of the extra factors but have good reasons for ignoring them here.

- **Propose different scenarios**: This is basically asking yourself 'what if…?' What if this factor changed? What if we took away this piece of equipment? What if we stopped selling the drug entirely? Look at the scenario and see how it would impact your current problem or situation. How would things change? Why?

- **Extreme extrapolation**: This is the more ridiculous version of the different scenarios from above. If you take the underlying factors which are most important in your current situation, what would happen if you pushed them to extreme, highly unlikely values?

- **Use a different framework**: Many management consultants are trained to use 'frameworks', essentially a pre-built way of looking at a situation with corresponding solutions to check. These are extremely helpful in many cases but may lead to you becoming stuck in one interpretation. Try using different frameworks, even ones which don't seem to apply completely. Do they bring up alternate possibilities?

Identifying relationships between elements

One of the most basic parts of solving any real-world problem is the ability to identify connections between apparently unconnected elements. This can be really, *really* hard, particularly in complex situations where many different factors are at play and the data you are working with is limited.

So here are a couple of tips to help out:

- **Look for similarities and differences**: Well duh, right? This is, unsurprisingly, the most important method. Look for the common themes or issues in the situation. Why are certain events happening in some cases but not others? Are there *outliers,* cases where the standard situation is not actually occurring? Why are they different? What factors separate them from the typical case?
- **Relative size**: Are some elements larger or smaller than others? Are some elements having a disproportionally large (or small) effect on the situation if you control for their relative size? If so, then why?
- **Cause and effect**: Do some elements cause other ones to occur? Is there a chain of events? Can you map these chains out, and if so is there a set of 'root causes' which the chain of events builds back to.
- **Sequence**: Inevitably some of your elements will happen before others, even if there does not appear to be a set cause-and-effect. Can you determine why this is occurring? Is there actually an underlying connection which may not be obvious at first sight?
- **Analogies**: Have you ever experienced or heard of a similar situation? Can you find a similar situation in your research? How did this relate to the current situation? What are the similarities? What was the ultimate solution for the analogous situation?

Identifying the heart of the situation

At the core of any complex problem is usually a single element or root cause which leads to the other elements. Finding this root cause is usually the only way to find a long-term solution, as solutions which target the secondary elements will generally stop working after a while.

What are some methods which can be used to find the heart of the situation?

- **Subconscious summaries**: *Quick,right now, summarise the problem in a single sentence*. This is an attempt to force your subconscious knowledge of the problem into the open, and will often provide you with a surprisingly good description of what is wrong. It is quite risky, however, as you may be lacking a piece of information which you need to find the true root cause rather than an 'almost true' one.
- **Analogies**: Have you come across a similar situation in the past or during your research? Was there a central point which was important to that situation? How does it relate to this one? Many problems are simply slight variations on previous problems – indeed most of the value of a consultant comes from their experience in previously solving similar problems.
- **Dis-aggregate**: Can you break up the situation? Are there sub-factors or sub-elements which seem more important than others? Do these have a common link?
- **Relative importance**: Are some elements more important than others? Can you rank them by their importance? Is the top one really the most important? Why?
- **Combination**: Can you combine the different elements to make an interesting story? Can you start with a recommendation and then arrange the facts to support this? Can you repeat this for multiple different recommendations? This is naturally dangerous, you need to avoid cherry-picking data to support a position, but can often help you to see the common thread in the situation.

- **Narrow the scope**: Can you examine just one piece of the situation at a time? Or implement a solution at one location prior to others? How will that help?Defining the problem

APPENDIX 4: CHECK-LIST FOR REVIEWING BRIEFING BOOKS

Early draft

There are several questions which you should ask as a reviewer of an early-draft briefing book:

- Is this document consistent with the high-level strategy which we are working towards? (As defined in the strategy document, planning discussions, etc.)
- Does the document meet the intended purpose?
- What data will support the planned arguments?
- Should other arguments be made in addition to those planned?
- Will the data be displayed in an appropriate and understandable way?
- Does other data visualisation (tables, figures) need to be planned to help support the strategy?
- Is it clear what the question being asked is, and what the company position will be?
- When we present the data, are we intending to show the most important information first?
- Do we explicitly show the rationale behind key statements?

Late draft

At this point there is more detail in the document, and so the questions which need to be asked while reviewing change slightly:

- Are there gaps in the data or unresolved issues?
- Is there enough context given in the briefing book to understand the issue?

- Are any open issues covered with data and clear arguments?
- Is the document logical, with reasonable and believable arguments?
- Are any required regulatory pieces included?
- Thinking with your 'health authority hat' on, would this document be convincing?
- Are figures and tables understandable without referring to the text?
- Are figures and tables well-presented and easy to understand?
- Does each section or paragraph begin with the main message of that subsection?

Final draft

The final draft is the chance to catch any last-minute problems and complete the final polish:

- Have all the issues been fully addressed in a way which is consistent across the entire document?
- If you spotted any gaps in the argument before, are they now fixed?
- Is the discussion clear and convincing?
- Do any sections contradict each other?
- Are the tables and figures correctly made and properly labelled?
- Is the text as concise as possible?
- Have all the necessary references (to literature, health authority guidance, regulatory feedback, etc.) been included?
- Are the footnotes correct?
- Are there any obvious typos, layout errors, or mistakes?

APPENDIX 5: GLOSSARY

Applicant: The company, institution or person who is applying for a scientific advice meeting.

Biosimilar: A 'generic for a biological drug', these are not identical due to inherent variability when manufacturing biological molecules. Significantly more complex and expensive to develop than a standard small-molecule generic.

Briefing book: The document(s) which you provide to the health authority describing your project, asking questions about development, and laying out your preferred solution.

BsUFA: The Biosimilar User Fee Act sets out the requirements for the FDA to support biosimilar development (including timelines for interactions and reviews) and formalises the amount that they can charge applicants to do so.

Clinical Hold: When a clinical study is paused due to critical challenges, most often safety-related.

Complete Response Letter: A slightly-misleading term, this is what you receive from the FDA when your regulatory application has been rejected. It's a response, but not a good one.

Complex Product: An FDA definition for generics which are more complex than the standard. These can receive additional support in the form of scientific advice meetings.

Conclusion first: Refers to the practice when communicating of leading with the main insight at the start, then providing supporting data. Very important for focusing readers on the key messages.

Controlled correspondence: Is a pathway by which generic drug manufacturers or others in the related industries can make a written request to FDA for feedback.

EMA: The European Medicines Agency, the centralised regulatory body for the European Union.

Face-to-face: An advice meeting which is held in a location where all members are physically present.

FDA: The Food and Drug Administration, the regulatory body of the United States.

Formal Consultation: Terminology used by PMDA in Japan to refer to a scientific advice meeting.

Formal Meeting: Terminology used by the FDA to refer to scientific advice meetings which can be held between the agency and the applicant.

GDUFA: The Generic Drug User Fee Act formalises requirements for the FDA to support generic drug development, as well as the fees associated with applications.

Guidance document: An official document issued by a health authority which sets out their thinking on a particular issue. If a guidance document says that you must do things in one particular way, it is a pretty clear hint to do so.

Health authority: A generalist term for the regulatory authority responsible for approving your drug, medical device, etc.

Initial application: The first application you make requesting approval for your product. Think of a BLA in the US or an MAA in the European Union.

Issue Cards: A card or piece of paper which describes a potential gap in your argument and your aligned response when someone notices it.

NMPA: The National Medical Products Administration is the health authority relevant for medicine in China.

PDUFA: The Prescription Drug User Fee Act sets out the requirements for the FDA to support drug development (including timelines for interactions and reviews) and formalises the amount that they can charge applicants to do so.

Pharmacopoeia monograph: Very comprehensive collections of requirements for things such as performing analytical tests, demonstrating sufficient quality of chemicals, and so on. The most important ones are USP and Ph. Eur., but many other versions exist.

PMDA: The Pharmaceuticals and Medical Devices Agency is the main health authority relevant for medicine in Japan.

Preliminary feedback: The feedback given to you just prior to the scientific advice meeting by the health authority, allowing you to adjust your plans for the meeting itself.

Protocol assistance: Enhanced support provided by EMA to companies developing drugs for under-represented therapeutic areas.

Refuse to File: A rejection of your application or submission because it did not meet the requirements. Very, very embarrassing when it occurs.

Scientific advice meeting: A meeting held with a health authority to discuss a challenge with your development program.

Scientific evidence: What the reviewer is looking to see when evaluating your claims – a series of valid data which supports your position.

SME: A Subject Matter Expert is someone in the team who has deep knowledge of a particular area being discussed.

Stakeholder: Someone with an interest in the outcome of a discussion or decision.

Strategy document: Also known as a 'Seed document', this provides a written outline of the strategy you will follow during the scientific advice process.

Tailored advice meeting: Specialised advice meeting offered by EMA for specialised areas of drug development, most notably in biosimilar development.

Teleconference: Discussion with the health authority over a phone line, without a physical presence.

Videoconference: Same as a teleconference, but with video links available. Has mostly replaced teleconferences with the major health authorities.

Written response only: Health authority feedback provided as a written document, with no meeting occurring to discuss.

Appendix 6: Further Reading

Problem solving:

Bulletproof Problem Solving, Conn

The McKinsey Approach to Problem Solving, Davis et al

Communication:

How To Lie With Statistics, Huff

Say It With Charts, Zelazny

Say It with Presentations, Zelazny

Storytelling with Data, Knaflic

Targeted Regulatory Writing Techniques, Foote

The McKinsey Thought Process, Roche

The Pyramid Principle, Minto

Health Authority Interactions:

Formal Meetings Between the FDA and Sponsors or Applicants of PDUFA Products, FDA

Formal Meetings Between FDA and ANDA Applicants of Complex Products Under GDUFA, FDA

How FDA Advisory Committee Members Prepare and What Influences Them, McIntyre et al, Ther Innov Regul Sci

About the Authors

If you liked this book then please take the time to review it online, every piece of feedback is appreciated and helpful

Nathan Martinsberg

Originally from the sunny shores of Australia, Nathan has worked in pharmaceutical companies based in many parts of the world. A varied career has given him a wealth of knowledge regarding biotech development, including due diligence and licensing. His books compile those years of experience and are written to help up-and-coming experts in the pharmaceutical field to get their bearings.

CF Harrison

CF Harrison currently works in the beer-filled heart of Bavaria. With a PhD in biochemistry, he has worked in drug discovery, as a scientific consultant, and as a regulatory affairs manager for a major international pharmaceutical company.

After realising that his friends from academia had no idea how the pharmaceutical industry actually worked, he decided to help answer their questions once and for all. His books simplify the complex and jargon-filled world of big pharma into useful information that is accessible for all.

Those interested in contacting him can drop a line to lifeafterlifescience@harrison-scientific.com.

OTHER BOOKS BY THE AUTHORS

Scientific Advice Meetings: A Guide to Successful Interactions with FDA, EMA and Beyond

Nathan Martinsberg and CF Harrison, available in eBook, Paperback and Hardcover

Scientific due diligence: A handbook for investigators and investors

Nathan Martinsberg, available in eBook and Paperback and Hardcover

Starting out in the pharma industry: Essential knowledge for life scientists

CF Harrison, available in eBook and Paperback

Pharmaceutical Regulatory Affairs: An Introduction for Life Scientists

CF Harrison, available in eBook and Paperback

Aseptic Production: An Introduction for Life Scientists

CF Harrison, available in eBook and Paperback

From test tubes to tonnes: Commercial drug process development for Life Scientists

CF Harrison, available in eBook and Paperback

INDEX